Я. Перельман

Занимательная физика

Книга 2

Издательство «Наука»
Москва

Ya. Perelman

Physics for Entertainment

Book Two

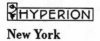

New York

Originally published in Russia by Mir Publishers
English translation © 1975, 2008 Mir Publishers

Published by Hyperion in the United States and Canada by arrangement with
Mir Publishers

This is Volume 2 of the original *Physics for Entertainment*. Both volumes were out
of print at the time of this publication and Volume 1 wasn't available, therefore
Volume 2 appears as a stand alone volume.

Library of Congress Cataloging-in-Publication Data is available upon request

ISBN: 978-1-4013-0921-3

Hyperion books are available for special promotions, premiums, or corporate
training. For details contact Michael Rentas, Proprietary Markets, Hyperion,
77 West 66th Street, 12th floor, New York, New York 10023,
or call 212-456-0133.

FIRST HYPERION EDITION

1 3 5 7 9 10 8 6 4 2

PUBLISHERS' NOTE

Perelman's *Physics for Entertainment*—this is a translation from the eighteenth Russian edition—owes its wide popularity to the rare talent of its author who was able to single out and present in an entertaining form ordinary facts and phenomena of profound meaning from the angle of physics. Perelman had a very definite purpose in mind when he wrote this book. Describing established conceptions and long known laws, he introduces us to the fundamentals of modern physics and tries to get us "think in physical categories". No wonder there is nothing in the book about the latest achievements in electronics, nuclear physics and the like. Though he wrote the book almost fifty years ago, he continually revised and supplemented it up to its thirteenth edition in 1936. In 1942 he died in the Leningrad blockade and subsequent editions were published posthumously.

We have not attempted to rewrite the book in this present edition, but have merely endeavoured to bring it up to date.

Contents

Chapter Three. Rotation

Chapter Four. Gravitation

Chapter Five. Travelling in a Projectile

Chapter Six. Properties of Liquids and Gases

Chapter Seven. Heat

Chapter Nine. Reflection and Refraction of Light. Vision

11

Chapter Ten. Sound. Wave Motion

1 *Fundamentals of Mechanics*

The Cheapest Way of Travelling

In his satirical *History of Lunar States and Empires* (1657) the witty 17th-century French writer Cyrano de Bergerac describes an amazing thing which had supposedly happened to him. Experimenting one day, he was lifted up into the air with all his retorts. On landing several hours later, he was astonished to find himself not in his own land of France nor even in Europe, but in Canada. Strangely enough Cyrano de Bergerac believed his transatlantic flight quite possible, claiming that while he was up in the air, the earth had continued to rotate eastwards which was why he had landed in North America and not France.

A very cheap and simple mode of travel, I must say! Just ascend and stay suspended a few minutes and you'll return to a totally different place much further westwards. Why tire yourself globe-trotting? Simply hover in mid air and wait till your destination reaches you.

Alas, this is nothing but a figment of the imagination. In the first place when we ascend in the air,

Fig. 1

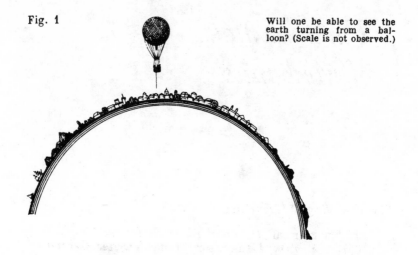

we don't really separate ourselves from Mother Earth. We are still tied together, because we are hanging in that envelope of air which also participates in the earth's axial rotation. The air—or rather its lower denser strata—spins together with the planet, carrying along everything in it—clouds, aircraft, birds and insects. After all if the air didn't spin together with our planet, we would always be buffeted by a wind of so terrible a force that in comparison with it the worst of hurricanes would seem a gentle zephyr (a hurricane, or tornado, moves with a speed of 40 m/sec or 144 km/h; at Leningrad's latitude, for instance, the earth would carry us through the air with a speed of 230 m/sec or 828 km/h). It would make no difference at all, whether we would be standing still with the air moving by, or whether the air would be still while we would be moving in it. In both cases we would feel the same strong wind. A motor cyclist dashing along with a speed of 100 km/h braves a formidable oncoming wind even in the calmest weather.

Then even if we were able to ascend to the top of the atmosphere or if the earth had no envelope of air at all, we would not be able to benefit by the cheap method of travel which the French satirist imagined. Indeed, when we separate from the surface of the spinning earth, *we continue by force of inertia to move with the same speed*—that is, the speed with which the earth is moving beneath us. Back on earth again we would find ourselves where we were before we went up. This is the same as to make a hop inside the carriage of a moving train. We hop up and land again in the same place. True, we would be moving by force of inertia rectilinearly (along a tangent), while the earth beneath us would be tracing an arc. For small intervals of time, however, this can be totally ignored.

Stop, Earth!

The celebrated British science-fiction novelist H.G. Wells tells the story of an office clerk who was able to work miracles. Though a rather dull young man, fate had endowed him with the surprising gift of making any wish that he expressed come true at once. However, this fascinating ability, it turned out, brought only trouble. As far as we are concerned, it is the end of the story that is instructive.

After a carousal, the clerk, scared of what his family would say if he turned up home in the wee hours of the morning, thought he might as well take advantage of his talent to prolong the night. Wondering how to do it, he decided to order the stars to stop in their tracks. He couldn't bring himself to do it, and when his friend suggested that he halt the moon, he took a long look at it and said musingly:

"'That's a bit tall.'

"'Why not?' said Mr. Maydig. 'Of course it doesn't stop. You stop the rotation of the earth, you know.... It isn't as if we were doing harm.'

"'H'm!' said Mr. Fotheringay. 'Well.' He sighed. 'I'll try. Here—'

"He buttoned up his jacket and addressed himself to the habitable globe, with as good an assumption of confidence as lay in his power. 'Just stop rotating, will you?' said Mr. Fotheringay.

"Incontinently he was flying head over heels through the air at the rate of dozens of miles a minute. In spite of the innumerable circles he was describing per second, he thought and willed: 'Let me come down safe and sound. Whatever else happens, let me down safe and sound.'

"He willed it only just in time.... He came down with a forcible but by no means injurious bump in what appeared to be a mound of fresh-turned earth. A large mass of metal and masonry ... ricochetted over him, and flew into stonework, brick, and masonry, like a bursting bomb. A hurtling cow hit one of the larger blocks and smashed like an egg.... A vast wind roared throughout earth and heaven, so that he could scarcely lift his head to look....

"'Lord!' gasped Mr. Fotheringay, scarce able to speak for the gale, 'I've had a squeak! What's gone wrong? Storms and thunder. ... It's Maydig set me on to this sort of thing. ...'

"He looked about him so far as his flapping jacket would permit. ... 'The sky's all right anyhow,' said Mr. Fotheringay. ... 'There's the moon overhead. But as for the rest—Where's the village? Where's—where's anything? And what on earth set this wind a-blowing? I didn't order no wind.'

"Mr. Fotheringay struggled to get to his feet in vain, and after one failure, remained on all fours, holding on. He surveyed the moonlit world to leeward, with the tail of his jacket streaming over his head. 'There's something seriously wrong,' said Mr. Fotheringay. 'And what it is—goodness knows. ...'

"Mr. Fotheringay ... perceived that his miracle had miscarried, and with that a great disgust of miracles came upon him. He was in darkness now, for the clouds had swept together and blotted out his momentary glimpse of the moon, and the air was full of fitful struggling tortured wraiths of hail. A great roaring of wind and waters filled earth and sky, and, peering under his hand through the dust and sleet to windward, he saw by the play of the lightning a vast wall of water pouring towards him....

"'Stop!' cried Mr. Fotheringay to the advancing water. 'Oh, for goodness' sake, stop!'

"'Just a moment,' said Mr. Fotheringay to the lightnings and thunder. 'Stop. ...'

"He remained on all fours ... very intent to have everything right.

"'Ah!' he said. 'Let nothing what I'm going to order happen until I say *Off*. ...'

"'Now then!—here goes! Mind about that what I said just now. In the first place, when all I've got to say is done, let me lose my miraculous power, let my will become just like anybody else's will, and all these dangerous miracles be stopped...'

"'And the second is ... let everything be just as it was before the miracles began. ... No more miracles, everything as it was—me back in the *Long Dragon* just before I drank my half-pint....'"

17

Imagine yourself in an airplane high up in the sky. You look down and see familiar places—you are approaching your friend's house. You think it wouldn't be a bad idea to send him a message. You quickly jot down a few words on your writing-pad, tear out the sheet of paper, wrap it around some heavy object—which for convenience's sake we shall henceforth call "weight"—and drop it as soon as your friend's house is right underneath. If you think it will fall into your friend's front garden, you're making a terrific mistake. You'll miss it as sure as eggs is eggs even though your friend's house is right below.

If you watched the weight as it fell you would see a strange thing. While falling, the weight at the same time will continue to *travel along beneath the plane*, as if tied to it by an invisible piece of thread. And as it falls on the ground it will be off the target by a long shot.

This is again a manifestation of that selfsame law of inertia, which prevents us from travelling in Bergerac's way. While the weight was in the plane it was moving together with it. But when dropped and having separated from the plane, it does not lose its initial speed. As it falls it continues to move in the air in the same direction as the plane. Both movements, perpendicular and horizontal, are added and, as a result, the load traces a curved trajectory, which keeps it beneath the plane—provided, of course, that the aircraft does not veer away from the original course or fly faster. In point of fact, the weight follows the same trajectory as that of a horizontally thrown body: a bullet discharged from a horizontally pointed rifle, for instance, would trace an arc-shaped trajectory that ends on the ground.

Fig. 2

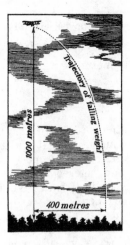

1000 metres

Trajectory of falling weight

400 metres

A weight dropped from an airplane in motion will fall not vertically down, but along a curved trajectory.

Note that all I have mentioned above would be valid if not for the air drag. Actually it impedes both the vertical and horizontal movements, the result being that the weight gradually lags behind the plane.

The deviation from a plumb-line trajectory may be pretty great when the plane is high up and is flying fast. On a windless day, a weight dropped from a plane flying with a speed of 100 km/h at an altitude of 1,000 metres would land some 400 metres in front of the spot that was directly beneath the plane when the weight was dropped (Fig. 2). We shall find the answer to the problem easily enough—providing, of course, that we ignore the air drag. The formula for computing the length of the path covered in the case of uniformly accelerated motion is $S=\frac{gt^2}{2}$, whence $t=\sqrt{\frac{2S}{g}}$. This means that from 1,000 metres up, it should take a stone $\sqrt{\dfrac{2\times1000}{9.8}}$ or 14 seconds to fall. In this period of time it will move horizontally forward by $\dfrac{100,000}{3,600}\times14=390$ metres.

Non-Stop Railway

If you were to be standing on an ordinary railway platform you would, of course, find it quite a feat to jump on a passing express. But suppose the platform would be moving, and, moreover, just as fast and in the same direction as the train. Would it be difficult for you to hop on then?

Not at all. You would be able to get on just as easily as you would when boarding a standing train. Once you and the train would be moving in the same direction with the same speed, *in its relation to you, the train would be in a state of rest.* Its wheels would be turning, true, but as far as you would be concerned, they would seem to be marking time.

Strictly speaking, all objects that we usually take to be standing still—as, for instance, a train that has come to a full halt at railway station—are really moving together with us around the earth's axis and around the sun as well. But since this motion doesn't bother us in the least, we can disregard it.

To all practical intents, we could easily get a train to take on and disgorge passengers without stopping. Exhibitions and fairs often provide such arrangements to enable visitors to see quickly and conveniently all there is to see. The entrance and exit of the fair ground are linked together by a non-stop railway: passengers may board, or alight from its moving cars at their convenience.

Fig. 3

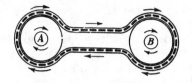

Diagram of a non-stop railway between stations A and B. The next figure shows how it works.

Figs. 3 and 4 give some idea of this interesting arrangement. *A* and *B* in Fig. 3 designate the termini. Each terminus has a circular *stationary platform* in the middle of a large rotating disc. Looped around the spinning discs of the two stations is a string of railway cars. Now see what happens when the disc revolves. The cars go round the discs with the same speed as that of the outer rim of the disc. Consequently, a passenger may safely hop on or off the train from the disc. After he alights, the passenger walks towards the centre of the rotating disc to the stationary platform in the middle. Here it is easy enough to cross from the inner rim of the moving disc to the stationary platform, since when the radius is small, circumferential velocity is also small (as is only natural, the points on the inner rim move much more slowly than the points on the outer rim, because in one and the same period of time they describe a far smaller circumference). Now the passenger has only to cross

Fig. 4

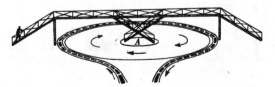

A station of a non-stop railway.

the overhead bridge to get to the ground outside the railway (Fig. 4).

Since there are no frequent stops much time and locomotive energy is saved. It is a fact that trams, for instance, spend most of the time and nearly two-thirds of the locomotive power to work up speed after stops and to slow down when coming to a stop.

Incidentally, the amount of energy lost when slowing down could be saved by causing the tram's electric motors to operate as dynamos and reverse current to the network. In this way the electricity expenditure on tramcar traffic in the Berlin suburb of Scharlottenburg was reduced by 30 per cent*.

Railway stations could even dispense with a special moving platform, to have trains take and disgorge passengers on the go. Imagine an express train dashing by an ordinary stationary platform. We want it to take on some more passengers without stopping. To do that these passengers should enter another train standing on a parallel reserve track. This train would start moving till it worked up the same speed as that of the express train. When the two trains draw parallel *they will be in a state of rest with respect to each other*. The passengers could easily get on the express by crossing gangways from the auxiliary train. Then there would be no need for trains to stop at stations.

Moving Pavements

Another device, so far used only at exhibitions—the "moving pavement"—is also based on the principle of the relativity of motion. The first moving pave-

---

* This method is now extensively employed on the electric Vladivostok-Moscow line.—*Ed.*

ments appeared at the 1893 Chicago Fair. The 1900 Paris Fair also had moving pavements.

Fig. 5 shows a nest of five pavement strips moving at different speeds. The outermost one is the slowest. Its speed is only 5 km/h, which being our ordinary walking speed makes it simple enough for us to get

Fig. 5

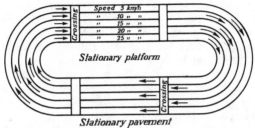

Moving pavements.

onto it. The second strip next to it already has a speed of 10 km/h. If we had to hop onto it from a stationary pavement we would find it pretty danger-ous, but to cross over to it from the first strip, is simple enough, because in relation to this first strip, with its speed of 5 km/h, the *second*, with its speed of 10 km/h, is moving with a speed of only 5 km/h. This means that it is just as easy to cross from the first strip to the second strip as it was to cross from the ground to the first strip. The *third* pavement strip has a speed of 15 km/h, but again it is easy enough to cross over onto it from the *second* strip. So will it be just as easy to cross from the third strip to the *fourth* one with its speed of 20 km/h, and finally from the fourth one to the *fifth* one, which already does 25 km/h. Standing on this fifth strip the pas-senger reaches his destination where he skips from strip to strip to get back to firm ground again.

None of Newton's three fundamental laws of dynamics is so perplexing as the famous third one—the *law of action and reaction*. Everybody knows it and some even know how to apply it correctly in certain cases. However few understand it fully. You may have been lucky enough to grasp its meaning at once, but I confess, for one, that it took me ten years before I got to the heart of the matter.

Most people with whom I have discussed this law are prepared to admit that it is right, making however a few essential reservations. They willingly admit that it holds for stationary objects but cannot understand how it applies to the interaction of moving bodies. According to the law for every action there is always an equal and opposite reaction. Consequently, when a horse pulls at a cart, it means that the cart is pulling at the horse with the same force. In that case the cart should stay where it is, shouldn't it? Nevertheless it moves. Why don't these forces offset each other, since they are equal?

That is the usual argument raised when this law comes up. Does this mean that the law is wrong? Of course not. It is just that we don't understand it correctly. The forces do not offset each other simply because they are applied to *different* bodies: one is applied to the cart and the other to the horse. The forces are certainly equal; but do equal forces always produce the same action? Do equal forces impart an equal acceleration to all bodies? Does not the action of a force on a body depend on the body itself? And on the value of the "reaction" which the body offers to the force? Once you think about it you will realise immediately why the horse pulls the cart along even though the cart is pulling the horse back with the same force. The force acting on the cart and the force acting on the horse are of equal magnitude at every moment, but since the cart freely moves on its wheels, while the horse pushes away from the ground, the cart rolls in the direction in which the horse is pulling it. Furthermore one must realise that if the cart did not "react" to the horse's motive power we would be able to dispense with the horse entirely, as the slightest push would already start the cart rolling. We need the horse to overcome the cart's reaction.

Perhaps you would grasp this point more easily were the law expressed not so laconically as it usually is—"action is equal to reaction"—but as: "The force of the reacting body is equal to the force of the acting body." After all it is only the *forces* that are of equal magnitude: the actions of the forces if understood—as they are usually understood—as the translation of a body, are, on the other hand, different as a rule, because the forces are applied to *different* bodies.

In February 1934 the Soviet ship *Chelyuskin* was crushed in the Arctic. Newton's third law easily explains why. When the ice pressed on the *Chelyuskin*'s hull, the hull pressed back with an equal force. The disaster occurred precisely because while the thick ice was able to withstand this pressure without crumbling, the hollow hull succumbed to this force and was crushed even though it was made of steel (you will find more about the *Chelyuskin* disaster further on).

Even in falling, every body strictly obeys the law of reaction. An apple falls, because it is attracted by the earth's gravity. However, *the apple itself attracts the whole planet with exactly the same force.* Strictly speaking, the apple and the earth fall towards each other, but their speeds of falling are different. The equal forces of mutual attraction impart to the

25

apple an acceleration of 10 m/sec² while to earth they impart an acceleration which is as many times less as many times the earth's mass is more than the mass of the apple. Naturally, the earth's mass is an incredible number of times greater than that of the apple. No wonder the earth's movement is so infinitesimally small that for all practical purposes it can be considered as nonexistent. Now you know why we say that the apple falls on the earth, instead of saying that the "apple and the earth fall on each other".

Why Did Svyatogor Perish?

Among Russian folk legends there is one about Svyatogor, an epic hero who tried to lift up the earth. According to another legend, Archimedes planned to do the same. All he needed, it is claimed, was a fulcrum for his lever. Svyatogor, however, possessed enormous strength and needed no lever. All he wanted was something he could grasp with his mighty hands.

"Could I gain a hold, I would lift the world."
Svyatogor dismounts from his trusty steed,
Takes hold of the bag with both his hands,
Then he raises it just above his knees,
While not tears, but drops of blood run a-down his face.
And he sank in the earth and could not get out,
And then it was that he met his end.

If Svyatogor had known the law of action and reaction he would have realised that his strength, when applied to the earth, would evoke an equal and, consequently, just as enormous opposite force which would draw him down into the ground. At any rate the legend shows us that the reaction the earth presents when pressed against had been obser-

26

ved long ago. People unconsciously applied the law of reaction thousands of years before Newton first enunciated it in his immortal *Principia*.

Can One Walk Without Support?

When we walk we push off with our feet from the ground or floor. We can hardly walk at all across a very smooth floor or ice, from which we can't push off our feet. A steam engine pushes away from the track with its driving wheels. But if we were to grease the rails, our locomotive would stay where it was. Sometimes, in icy weather, the rails are sanded in front of the locomotive's driving wheels to get the train started. When wheels and track—at the time railways first appeared—were geared to each other, the idea was that the wheels should push themselves away from the rail. A ship pushes itself away from the water by means of its paddles or screw. A plane also pushes itself away from the air, with the aid of a propeller.

To make a long story short, in whatever medium an object moves, it uses it as a support when moving. But would a body be able to move, if it *didn't have any support*?

You would think this incredible, wouldn't you? It would be like lifting oneself up by one's hair and that was something which only that king of liars, Baron Munchausen, could do. Nevertheless we often see this seemingly impossible motion. It is true enough that a body cannot start moving all by itself, due to the effort of inner forces alone. But it can make part of itself move in one direction and the rest in the opposite direction. You have probably seen a rocket whizzing up into the air. But have you ever stopped to wonder why it shoots up? It provides a most graphic illustration of the kind of motion we are now discussing.

Why Does a Rocket Go up?

Even from students of physics one may often hear a totally wrong explanation of a rocket's flight. They claim that it goes up by *thrusting itself away from the air* with the help of the gases formed from gunpowder combustion. That incidentally is what ancients thought—rockets were invented long, long ago. But if we were to fire a rocket in an airless void it would fly, and even better than in the air. The real cause of a rocket's flight is absolutely different.

The Russian revolutionary Kibalchich, who was executed for attempting to assassinate Tsar Alexander II, furnished a very lucid and simple exposition of rocket motion in the notes he wrote in his death cell and in which he described the flying vehicle that he had invented. Explaining the design of a rocket which was to be used as a military weapon he wrote: "In a tin cylinder, closed at one end and open at the other, we insert a cylinder of the same size consisting of closely packed gunpowder with a channel in the centre. Combustion begins at the surface of this channel and spreads within a definite period of time to the outer surface of the packed gunpowder. The combustion gases press on every side: but while the pressure the gases exert sideways is offset, the pressure they exert on the bottom of the tin cylinder is not offset by an opposite pressure—as here the gases have a free outlet. They are thus able to thrust the rocket forward in the direction in which it was mounted before ignition."

The same thing takes place when a projectile is

fired from a gun—the projectile flies forward, while the gun rolls back. Take the "recoil" of a rifle, or of any firing weapon for that matter. If our gun were suspended in mid air and had nothing to rest upon, it would move back, after the projectile were fired, with a speed as many times less than that of the projectile as the latter would be lighter than the gun itself.

In Jules Verne's science-fiction novel *Upside Down*, its heroes even thought of using the recoil of a tremendous cannon to carry out the stupendous project of "straightening the earth's axis".

The rocket is also a gun, but instead of shooting out projectiles it gives a burst of combustion gases. It is this that explains the spinning of the so-called Catherine wheel which you may have seen during firework displays. When the powder starts burning in the squibs attached to the wheel, the combustion gases stream out at one end, while the squibs themselves and the wheel to which they are attached revolve in the opposite direction. Actually, this is merely a modification of that well-known physical appliance called Segner's wheel.

Curiously enough, before the steamship was invented, there was a project for a mechanically-driven ship based on the same principle. The idea was to eject a jet of water through a powerful pump mounted in the stern, thus sending the ship forward, in the same manner as those floating pieces of tin, which school physics labs employ to demonstrate the principle under consideration. This project was not realised at the time, but it helped Fulton to invent his steamship.

We also know that the oldest steam engine, which Heron of Alexandria invented way back in the second century, was based on the same principle. The steam from the boiler (Fig. 6) travelled along a pipe into a sphere mounted on a horizontal axle. Spurting out of elbow pipes, the steam thrust these pipes in the opposite direction and the sphere began to revolve. Unfortunately Heron's steam turbine remained just a curious toy, because since slave labour was so cheap in ancient times nobody ever thought of deriving any

Fig. 6

World's oldest steam machine or turbine, which legend has ascribed to Heron of Alexandria (circa 200).

practical benefit from it. However its principle has not been forgotten; today it is applied in the making of jet turbines.

One of the earliest designs for a steam-driven automobile based on the same principle is ascribed to Newton, the author of the law of action and reaction. According to it the steam coming from a wheel-mounted boiler spurted out in one direction to make the boiler itself recoil in the opposite direction (Fig. 7).

Rocket automobiles are a modern modification of Newton's carriage.

For those who like to make things, Fig. 8 shows a paper ship resembling Newton's carriage. It has a steam boiler made of an empty egg-shell, which is

Fig. 7

Steam-driven automobile, supposedly invented by Newton.

Fig. 8

Toy paper steamship with egg-shell boiler. Heat is provided by a thimbleful of alcohol. The steam the egg-shell boiler ejects compels the ship to move contrariwise.

heated by means of an ignited piece of cotton wool soaked in alcohol and placed in a thimble below the egg-shell. The burst of steam from the shell sends the "ship" in the opposite direction. One must be very good with one's hands to make this most instructive toy.

How Does a Cuttle-Fish Swim?

It may strike you as odd to hear that there are quite a few creatures for whom the "lifting up of themselves by the hair" is quite a usual way of swimming. The cuttle-fish and, in general, most *cephalopo-*

Fig. 9

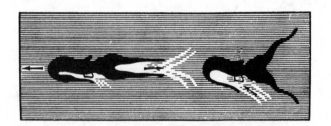

How the cuttle-fish swims

dae propel themselves through water in this fashion. They draw water into their gills through a slit in the side and a special funnel in front. Then they spurt out a jet of water through this funnel. This gives them, in conformity with the law of reaction, a backward impetus strong enough to cause the back of their body to move forward. The cuttle-fish, incidentally, can direct its funnels sideways or backwards and, by sending out jets of water, move in any direction.

32

The jelly-fish moves in the same fashion: contracting its muscles, it ejects the water from beneath its umbrella-shaped body, thus receiving a recoil. *Salpas*, the larvae of *dragon-flies* and some other aquatic creatures swim in the same fashion. Yet we doubted it!

Rocket Travels to the Stars*

Could anything be more thrilling than to travel to the moon and from planet to planet? How many science-fiction novels have been written on that subject. Voltaire in *Micromégas*, Jules Verne in *A Journey to the Moon* and *Hector Servadac* and H.G. Wells in *The First Men in the Moon* have, together with their many less-talented colleagues, taken us on many an exciting journey to the celestial objects, in imagination, of course, because we are still captives of our home planet.

Couldn't we make this dream that man has cherished for ages come true? Will all the clever projects that the science-fiction novelists have suggested and that are all so much like the real thing never be accomplished?

We shall go back to fantastic projects for interplanetary travel, later. Meanwhile let me tell you about

---

* Today, when artificial earth satellites go up with astonishing regularity and space probes reach neighbouring planets, when man has set foot on the moon and samples of lunar rock have been brought back to earth, our younger reader—who will no doubt take all this more or less for grunted as he or she will have been conscious from early childhood, if not the cradle, of the age of space exploration, which began in 1957— may think the author's glowing enthusiasm for space travel somewhat naïve. Still we have decided to reproduce the related passages as written, as they are of definite historical interest.—*Ed.*

33

a quite feasible project which was first suggested by the Russian scientist Konstantin Tsiolkovsky.

Could one fly to the moon in an airplane? Of course not. After all planes and dirigibles fly solely because they float on the air and thrust themselves away from it. There is no air between earth and moon and, in general, nothing dense enough to support an "interplanetary dirigible". Consequently, one has to invent a vehicle that would be able to fly without any support. We have already discussed a vehicle of this kind —the toy rocket. Couldn't we construct a huge rocket with special compartments for people, provisions, air tanks and all the other necessary things? Imagine the crew of such a rocket would carry with them a big enough store of fuel and be able to send out bursts of gas in any direction. This would be a real guided spaceship, one that would be able to take us to the moon and the planets. By firing these bursts of gas its crew would be able gradually to accelerate motion in such a way that would cause no harm. If they thought of landing on some planet they could, on the contrary, decelerate to make a soft landing. And they could employ the same method to return to earth.

It seems only recently that airplanes were just timidly getting into stride. Today they cross mountains, deserts, continents, and oceans. Couldn't we imagine interplanetary travel achieving similar progress in some 20 to 30 years from now? Then, at last, man will be able to break the invisible fetters that have chained him to this planet from times immemorial and strike out into the infinite reaches of the Universe!

Force. Work. Friction

The Krylov Fable Problem

The story of how the swan, the crawfish and the pike attempted to get a cart going is the subject of a well-known fable, one variant of which was written by the 19th-century Russian writer Ivan Krylov. I hardly suppose any one of you has ever stopped to examine it from the angle of mechanics. The result we would obtain in that case would be quite different **from** Krylov's denouement. The fable actually poses a problem in mechanics with several forces acting at angles to each other—the swan pulling the cart up, the crawfish, back, and the pike, into the river. Fig. 10 gives us these three forces—the swan's upward pull (*OA*), the pike's sideways pull (*OB*) and the crawfish's backward pull (*OC*). Don't forget that we have yet a fourth force—the cart's weight, which is directed downwards. Krylov claims in the fable that the cart remained where it was, or, in other words, that the resultant of all the forces applied to the cart is nil.

Is this really so? The swan pulling upwards is not in the way of the crawfish and the pike. On the con-

Fig. 10

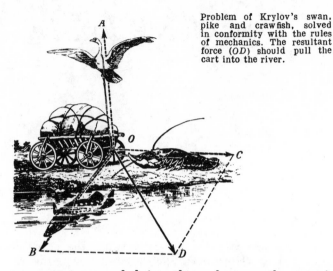

Problem of Krylov's swan, pike and crawfish, solved in conformity with the rules of mechanics. The resultant force (OD) should pull the cart into the river.

trary, it is even helping them, because the swan's pull, being directed against the earth's gravitational pull, lessens the friction between the wheels and the ground and between the wheels and their axles thus reducing the cart's weight and perhaps even offsetting it completely—since according to the fable the cart was rather light. Let us suppose for simplicity's sake, that the swan's pull does indeed offset the cart's weight. This means that we have only two forces left—the crawfish's pull and the pike's pull. From the fable we know the direction in which these forces are applied —the crawfish is pulling the cart backwards while the pike is pulling it into the water. It stands to reason that the river must have been to the side and not in front of the cart—because after all said and done Krylov's three toilers certainly never intended to topple the cart into the river. Consequently, the two pulls of the crawfish and the pike are set at angles to each other. Once the applied forces are not set in one and the same direction, the resultant cannot be nil.

Applying the rules of mechanics, we construct a parallelogram of forces along *OB* and *OC*, the diagonal of which, *OD*, gives the direction and magnitude of the resultant force. It is quite plain that this resultant force must cause the cart to move, all the more since its weight has been fully or partially offset by the swan's pull. The next question is: In which direction will the cart move—forward, backward, or sideways? This naturally depends on the ratio of the two forces and the value of the angle between them.

Those of you who have added and resolved forces before, will realise that even when the swan's pull does not offset the weight of the cart it cannot stay put. It won't move only if the friction between the wheels and the axles and between the wheels and the ground is greater than the applied forces. In that case the cart wouldn't have seemed so light—as Krylov claims. At any rate the poet had no ground to say that the cart would have stayed put—which of course does not change the moral of the fable.

In Defiance of Krylov

We have just seen that Krylov's moral, that "when friends fall out, nothing goes right" does not always dovetail with mechanics. Forces may not be applied in one direction, but notwithstanding will give some result. Few know that those assiduous toilers, the ants, whom the same Krylov praised as exemplary workmen, do their job by using the very method that he ridiculed, and do get things going. Again it is thanks to the composition of forces. If you take the trouble to watch ants at work, you will see that their supposedly intelligent cooperation is a fiction because actually it is a case of each ant for itself, without caring for what the others do.

This is how Elachich, the zoologist, described ants at work in his book *Instinct*:

"When you have some dozen ants dragging a large object across even ground, all act in the same fashion,

Fig. 11

Ants dragging a caterpillar.

and you have, on the face of it, what seems to be co-operation. Suppose, however, that the trophy, a caterpillar, for instance, is obstructed by a blade of grass or a pebble. The obstacle has to be circumvented: it is no longer possible to go on dragging the caterpillar forward. It is then that you clearly see how each ant tries to cope with the obstacle separately, by itself,

Fig. 12

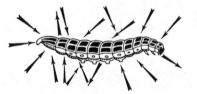

How ants work. The arrows show the approximate direction of the efforts of each ant.

without any thought of cooperating with its mates (Figs. 11 and 12). One pulls to the right, another to the left, a third forward and a forth backward. They change places, grasp the caterpillar in another place, each one pushing or pulling by itself. When, finally, it so happens that the forces of the ants are applied

38

in such a way that you have four ants moving the caterpillar in one direction and six in another, eventually it moves in the direction that the six are pulling, despite the resistance offered by the other four."

Let me give you another instructive instance to graphically illustrate this sham cooperation between ants. Fig. 13 shows a rectangular piece of cheese with 25 ants pulling at it. The piece of cheese slowly moves in the direction indicated by the arrow A. You may think that while the front row is pulling the cheese forward, the back row is pushing it also forward and the ants at the sides are helping their mates. But this is not at all so. Take a knife and separate the back row.

Fig. 13

How ants drag a piece of cheese to their ant-hill, to which arrow A points.

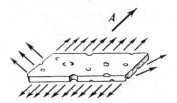

Immediately the piece of cheese begins to move forward much faster. This shows that the eleven ants in the back row were pulling the cheese backward and not pushing it forward. Each of them was plainly trying to deliver the piece of cheese to the ant-hill by pulling it backward, which clearly shows that far from helping the front row the back row was assiduously hampering them, countering their exertions. Actually the exertions of four ants would have been quite enough to pull the cheese forward, but since they do not coordinate their actions at all, it needs 25 ants to push the cheese forward.

Mark Twain happened to note this singular "cooperation" between ants. When describing an encounter

39

between two ants, one of which had chanced upon the leg of a grasshopper, he wrote:

"... they take hold of opposite ends of that grasshopper leg and begin to tug with all their might in opposite directions. ... They decide that something is wrong, they can't make out what. ... Mutual recriminations follow. ... They warm up, and the dispute ends in a fight. ... They make up and go to work again in the same old insane way, but the crippled ant is at a disadvantage; tug as he may, the other one drags off the booty and him at the end of it. Instead of giving up, he hangs on. ..." Though Twain was poking fun, he was absolutely right when he remarked that:

"He does not work, except when people are looking, and only then when the observer has a green naturalistic look and seems to be taking notes."

Crushing an Egg-Shell

One of the philosophical problems, which that "mighty intellect", Kifa Mokievich—a character from the novel *Dead Souls* by the great 19th-century Russian writer Gogol—racked his brains to solve was the following: "Suppose elephants were born in eggs. Wouldn't the egg-shell be very thick? I wager that even a cannon ball wouldn't pierce it and that a new firing weapon would have to be invented."

I'd wager that Gogol's philosopher would have gasped with amazement had he been told that the ordinary *egg-shell* is not so fragile as would seem. It's pretty hard to crush an egg between one's palms in the way shown in Fig. 14. You would have to exert quite an effort. (If you ever try it, beware of the shell splinters.)

Why is the egg-shell so strong? Solely because of its curved shape, which explains too the strength of all vaults and arches.

Fig. 14

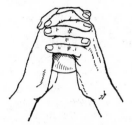

It takes quite an effort to break an egg this way.

Fig. 15 shows a small stone window arch. Load *S* (the weight of the masonry above it) pressing down on the wedge-shaped brick in the middle of the arch exerts a force designated by the arrow *A*. However the crowning brick can't fall down because of its wedge-shaped form; it merely presses against its neigh-

Fig. 15

Why an arch is so strong.

bours. Force *A* is thus resolved—according to the rule of the parallelogram—into two forces designated by the arrows *C* and *B*. These two forces are offset by the resistance offered by the adjacent bricks, which in turn are sandwiched between the others. Hence the force pressing down on the arch from above won't crush it. On the other hand, it is rather easy to ruin the arch by pressing up from inwards. No wonder,

41

because while the wedge-shaped form of the bricks prevents them from falling out *downwards*, it can't prevent them from being pushed *upwards*.

Our egg-shell is also an arch only one that is curved on every side. *Outside* pressure won't break it so easily as you might expect it to. You could stand an oak table by its four legs on four raw eggs and they will not be crushed (to make eggs stand you must mount them on supports of plaster-of-Paris, which easily adheres to the lime egg-shell).

Now you see why a sitting hen need not fear that the eggs will be crushed by the weight of its body. But the weakling chick inside easily shatters the egg-shell from within as it emerges from "Nature's dungeon".

When snapping off the top of the egg by striking at it sideways with a spoon, you can't believe that it so staunchly resists Nature's thrusts. Nature has certainly provided a fine "coat of mail" for the developing embryo inside!

The egg-shell also explains that mysterious strength of the seemingly fragile electric bulb—which is all the more amazing when one considers that bulbs have *nothing* inside to resist the quite substantial pressure of the outside air. You might be interested to learn that a 10-cm bulb has to resist a combined pressure of more than 75 kg—a full-sized man's weight. Incidentally, it has been experimentally demonstrated that a bulb can withstand a pressure that is even two and a half times greater.

Sailing "Close-Hauled"

How do sailing vessels manage to sail "to the wind", or as seamen say "close-hauled"? True, a seaman will

tell you that you can't sail directly against the wind, but

nearly against it, at an acute angle to the directio·
in which the wind is blowing. The angle however is
very small—about a quarter of a right angle—and
it is indeed hard to understand what difference there
can be between sailing directly against the wind or
at an angle of 22° to it.

Actually there is a difference and I shall explain
how a sailing vessel can get the force of the wind to
help it go "close-hauled". First, however, let us see

Fig. 16

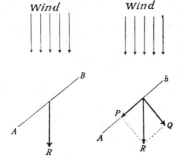

Wind will always push a
sail at right angles to its
plane.

how the wind acts on the sails, in general, or in other
words, in which direction it pushes the sail when
blowing against it. I suppose you think the wind
always pushes the sail in the direction in which it is
blowing. This is not so. Whatever the direction in
which the wind may be blowing it will always push
the sail in a direction perpendicular to the sail's
plane. Imagine the wind to be blowing in the direction indicated by the arrows in Fig. 16 in which the
line *AB* is the sail. Since the wind presses evenly over
the entire surface of the sail, we may replace the wind's
pressure by the force *R* as applied to the centre of the

sail. Resolving this force we get force Q, which is perpendicular to the sail, and force P, which is directed along it (Fig. 16, right). The latter does not push the sail at all as friction between the wind and the canvas is negligible. All we have left is force Q which pushes the sail at right angles to it.

Once we are aware of that we shall easily understand why a sailing vessel is able to go nearly against the wind at an acute angle to it. Let KK in Fig. 17 be the vessel's keel line. The wind is blowing at an acute angle to this line in the direction indicated by the arrows. AB is the sail itself, which is set so that its plane bisects the angle between the direction of the keel and the direction of the wind. Fig. 17 shows how the force

Wind

Fig. 17

How to sail near the wind.

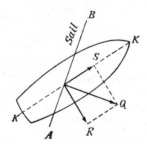

is resolved. The pressure the wind exerts on the sail is designated by force Q, which, as we know, must be perpendicular to the sail. Resolving this force we get force R, which is perpendicular to the keel and force S, which is directed forward along the boat's keel line. Since the boat's movement in the direction R encoun-

44

ters a strong resistance offered by the water (the keels of sailing vessels go very deep), force R is almost fully offset by the water's resistance. All we have left is

Fig. 18

Yacht tacking.

force S which, as you see, is directed forward and which, consequently, impels the boat forward at an angle, into the teeth of the wind, as it were (one could prove force S is greatest when the sail's plane bisects the angle between the directions of the keel and the wind). Usually this manoeuvre is executed by zigzagging in the way shown in Fig. 18—which in the seamen's lingo is called "tacking".

Could Archimedes Have Ever Moved the Earth?

"Give me where to stand and I will move the earth!" is a saying that legend has ascribed to Archimedes, the genius of antiquity who discovered the laws of the lever. "Archimedes," Plutarch says, "once wrote to King Hiero of Syracuse, whose kinsman and friend he was, that this force could be used to move any weight. Carried away by the power of argument, he added that, were there another earth, he would go there and lift our own planet from it."

Archimedes knew that by using a lever one could lift the heaviest of weights by applying even the weakest of forces. One had only to apply this force to the lever's longer arm and cause the shorter one to act on the load. He therefore thought that by pressing with his hand on the extremely long arm of a lever he would be able to lift a weight, the mass of which would be equivalent to that of the earth (for clarity's sake we shall take the "moving" or lifting of the earth to mean the lifting *on the earth's surface* of a weight whose mass would be equivalent to that of the earth).

I believe that if this great scholar of antiquity would have known what an enormous mass the earth possesses, he would have most likely "eaten his words" Imagine for a moment that Archimedes had at his disposal "another earth" and also the point of support he sought. Further imagine that he was even able to manufacture a lever of the required length. I wonder if you can guess the time he would need to lift a load equivalent in mass to that of the earth, by at least a centimetre? *Thirty million million years—* and no less!

Astronomers know the earth's mass. (See my *Astronomy for Entertainment** to learn how it was ascertained.) On earth a body possessing such a mass would weigh in round numbers

6,000,000,000,000,000,000,000 tons.

 * Foreign Languages Publishing House, Moscow, 1958.

Supposing a man could lift only 60 kg directly, to "lift the earth" he would need a lever with a long arm that would be longer than the shorter arm by

100,000,000,000,000,000,000,000 times!

You can easily figure it out that to have the end

Fig. 19

"Archimedes moving the earth". (Reproduction of an engraving from Varignon's book on mechanics [1787].)

of the short arm rise by one centimetre, the other end must describe through space the huge arc of

1,000,000,000,000,000,000 km.

That is the colossal distance Archimedes would have had to push the lever to lift the earth by just one centimetre. So how much time would he need? Presuming Archimedes could have lifted 60 kg one metre in one second—the work of almost one horse-power!—to lift the earth by just one centimetre, even then he would need

1,000,000,000,000,000,000,000 seconds

or 30 million million years. Though he lived to a ripe old age Archimedes and his lever wouldn't have lifted the earth by so much as even the thinnest of hairs.

No artifices would have helped him to cut the time noticeably—despite all his brilliance. For according to the "golden rule" of mechanics, the mechanical advantage derived will always be accompanied by a loss in displacement, or, in other words, in time. Even if Archimedes had been able to push the lever with a speed of 300,000 km/sec—the speed of light, and Nature's fastest—he would have lifted the earth by one centimetre only after *ten million years* of pushing.

Jules Verne's Strong Man and Euler's Formula

Do you remember Matifou, the Herculean character of one of Jules Verne's novels! He had a "magnificent head which was in good proportion to his giant stature. His chest was like smithy bellows, his legs like thick logs, and his hands like real cranes with fists that looked like hammers". One of his exploits described in the novel *Mathias Sandorf* is the amazing case of the good ship *Trabacolo*, which our giant kept in place with his brawny arms. This is how Jules Verne describes the exploit:

"The *Trabacolo* was about to be launched. Only a few wedges were left and some half a dozen carpenters were busy hammering away. Meanwhile a crowd of idlers looked on.

"At that moment a pleasure yacht shot out from beyond the promontory. Since it had to pass by the *Trabacolo* to reach the port, launching operations were temporarily suspended. For otherwise, had the two ships collided, the yacht would have gone down at once.

"All eyes were turned to the handsome vessel, whose white sails gleamed golden in the sunshine. It had just come across when a cry of horror rent the air. The *Trabacolo* shuddered and started to slide down the slips stem forwards.

"Suddenly a man leapt forwards, caught at the tow-lines and, in the twinkling of an eye, wound them round an iron stake driven into the ground. Running the risk of being crushed to pulp, he held on with a superhuman effort for some ten seconds before the tow-lines snapped. However this was enough, the *Trabacolo* barely grazed the yacht as it plunged into the water.

"The hero was none other than our old friend Matifou."

How astonished I imagine Jules Verne would have been had he known that one by no means has to be a giant with the "strength of a tiger" to do what Matifou did. Any resourceful person could do the same.

Mechanics tells us that a piece of rope wound around a drum produces when sliding a great deal of friction—which increases in geometric progression to the increase, in arithmetic progression, of the number of turns. This means that even a little child could hold a tremendous load by means of a rope wound three or four times around a post. At river ports teen agers use this method to bring to a halt boats with hundreds of passengers on board. It isn't superhuman strength, but friction that helps.

The famous 18th-century mathematician Euler established the proportion by which friction increases depending on the number of turns. For those undaunted by algebra's laconic lingo, here is Euler's most instructive formula, to wit: $F = fe^{k\alpha}$, where F is the force against which we direct our effort f, e is the natural logarithm base 2.718..., k is the coefficient of friction of rope on stake, and α is the "angle of turns", or the ratio between the length of the arc covered by the rope to its radius.

Applying this formula to Jules Verne's case, we get a staggering result. In the case in question F is the vessel's pull as it slides down the slip. From the novel we know that the ship weighed 50 tons. Presuming that the tilt of the slip was 1:10, it was not the ship's full weight that bore down on the rope but only a tenth of it—five tons or 5,000 kg. Let us take k, the coefficient of the rope's friction on the iron stake to be 1/3. It is easy to find α, since we know that Matifou wound the rope around the stake only three times. In that case

$$\alpha = \frac{3 \times 2\pi r}{r} = 6\pi.$$

Back now to Euler's formula to get the equation:

$$5,000 = f \times 2.72^{6\pi \times 1/3} = f \times 2.72^{2\pi}$$

whence f — the effort we must make — can be determined by using logarithms. Thus,

$$\log 5,000 = \log f + 2\pi \log 2.72$$

whence

$$f = 9.3 \text{ kg}$$

So to hold the ship the giant had to pull at the rope with a force of only 10 kg.

Now don't think the 10 kg mentioned merely a theoretical figure and that actually you would have to exert a much greater effort. On the contrary, the figure is even greater than it should be because when you have a *hemp rope* wound around a *wooden stake*, the coefficient of friction k is bigger and your effort f would be ridiculously small. All we must hope for is that the rope wouldn't snap from the strain. Then a little child, using a piece of the rope wound around a stake three or four times, could outvie Jules Verne's strong man.

Why Do Knots Hold?

In everyday life we often unsuspectingly derive advantages from *Euler's formula*. After all what is a knot if not a piece of twine wound around a small cylinder, the role of which in this case is played by another part of the same piece of twine? The strength of different knots, used by seamen or otherwise, depends exclusively on friction, which is enhanced several-fold by the string being wound around itself— just as the rope was wound around the stake. You can verify my statement by tracing the bends of a knotted string. The more bends, the greater the number of times the string is wound around itself, the greater the "angle of turns", and, consequently, the firmer the knot.

A tailor unconsciously applies the same principle when he sews on a button. He winds the thread many times around the stitch through the cloth and then snaps it off. As long as the thread is strong enough the button will hold on. Again the familiar rule applies: the number of turns increases in arithmetic progression, while the strength with which the button holds increases in geometric progression. In the absence of friction we wouldn't be able to have buttons on our clothes, their weight would cause the thread to unwind and they would drop off.

Supposing There Were No Friction

You have seen the varied, at times, unexpected ways in which friction exhibits itself. Friction, incidentally, is the star actor in cases where we don't even guess that it's there. If friction were suddenly to vanish, many things we are so accustomed to would go all awry.

The French physicist Guillaume has given us a very picturesque description of the role that friction plays:

"You have all happened to walk along icy pavements and no doubt remember how hard it was to keep your balance. How many funny jiggles you had to make! This makes us admit that the earth on which we live and walk usually has that precious property, thanks to which we keep steady without any particular effort. It's the same when we cycle along a slippery road or when a horse slips on asphalt and falls. It is by studying these things that we discover the consequences of friction. Engineers try the best to rid machines of it and they're doing the right thing. In applied mechanics friction is regarded as most undesirable—and again quite rightly ... but for a very narrow special field. In all other cases we must be grateful to friction. It enables us to walk, sit and work unafraid that books and inkpots would slip off onto the floor or that tables would slide until they bumped into a corner, or that pens would slip out of our fingers.

"Friction is so common that with the exception of a few rare cases we never have to invoke it: it comes of its own accord. Friction makes for stability. Carpenters level floors so that tables and chairs stay put. All the crockery and glassware we lay a table with, remain where they are without us having to worry about them—as long as we aren't on a ship tossing in a pitching sea. Now imagine we could get rid of friction completely. Then nothing—be it a huge slab of stone or a tiny grain of sand—would ever stay put. Everything would slide and roll until

all bodies reached one and the same level. If we had
no friction the earth would resemble a smooth ball,
the shape of a drop of water."

Add to this that in the absence of friction nails
and screws would slip out of walls, we wouldn't be
able to hold a thing, no whirlwind would ever stop,
and no sound would ever cease, continuing as an end-

Fig. 20

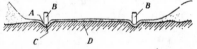

Top: A sledge on an ice-way:
two horses are pulling a
70-ton load. Bottom: the
ice-way: *A*—the track; *B*—
the runners; *C*—the packed
snow; *D*—the earth base.

less echo bouncing back from the walls of a room,
for instance, without growing any weaker.

Icy pavements show us how tremendously important
friction is. When we go out of doors in such weather,
we find ourselves in a helpless plight: we are always
afraid to fall. Here are some instructive newspaper
clippings for December 1927:

"London, 21. Due to very icy weather street and
tram traffic in London has been experiencing noticeable difficulties. Some 1,400 people have been hos-
pitalised with fractures."

"The petrol burst into flames and completely destroyed three cars which had collided with two tramcars near Hyde Park."

"Paris, 21. There have been many accidents in Paris and its suburbs due to icy weather."

However, the negligible friction that we get on ice finds a good technical application. The ordinary sledge is one example. A still better instance is afforded by the so-called ice-ways made to haul felled timber to the railway station or rafting place. On such a road (Fig. 20) with rails of smooth, slippery ice two horses can pull a sledge carrying as much as 70 tons.

Physical Causes of the *Chelyuskin* Disaster

I hope however, that I haven't made you jump to the conclusion that friction on ice is always negligible. Even at a near-zero temperature it can rather often be pretty great. A thorough study had been made of the friction of Arctic ice on the steel skin of icebreaker hulls. The coefficient proved to be unexpectedly big—0.2, as much as the friction of iron on iron. To realise the importance of this coefficient for icebreakers we shall examine Fig. 21, which shows the direction of the forces acting on the hull MN, when ice is pressing against it. Force P, the ice's pressure, is resolved into two forces: R, which is perpendicular to the hull, and F, which is directed at a tangent to the hull. The angle between P and R is equal to the angle α of the side's inclination to the vertical plane. Force Q, the friction of ice on the hull, is equal to R multiplied by the coefficient of friction, 0.2, meaning that $Q = 0.2R$. When Q is smaller than F, the latter pulls the pressing ice under the water, with the result that the ice slides alongside the hull without doing any damage. But should Q be greater than F,

Fig. 21

Icebound *Chelyuskin*. Bottom. Forces that act on the vessel's hull *MN*, when ice presses against it.

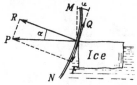

friction will prevent the ice from sliding alongside the hull and after some time the ice may dent and even crush the hull. When is Q less than F? F is clearly equal to $R \tan \alpha$, consequently, we should have the inequality: $Q < R \tan \alpha$; and since $Q = 0.2R$, the inequality $Q < F$ leads us to another inequality:

$$0.2\,R < R \tan \alpha$$

or

$$\tan \alpha > 0.2.$$

Now we use the tables to find the angle, the tangent of which is 0.2: it is 11°. Consequently, Q is less than F when α is greater than 11°. This tells us what inclination of the sides to the vertical plane guarantees safe progress through ice. It must not be less than 11°.

Now back to the *Chelyuskin* disaster. This ship—it wasn't an icebreaker—successfully voyaged the entire

Northern Sea Route but was jammed by ice in Bering Strait. The drifting ice carried the *Chelyuskin* northwards and finally crushed it (in February 1934). Many may still remember the two-months *Chelyuskin* Odyssey and the rescue of its crew by Soviet airmen.

Here is a description of the disaster itself. "The strong plating did not yield at once," expedition chief Otto Schmidt reported by radio. "We could see the ice pressing into the hull and the plates above it bulging. The ice pressed on, slowly but inexorably. The steel plates burst at the seam and all the rivets flew out. In an instant the port side was peeled off from prow to stern."

By now you ought to understand what caused the disaster. The conclusion is that the sides of ships built for voyaging in icy seas must have a minimum inclination of 11°.

Self-Balancing Stick

Balance a smooth stick on your outstretched forefingers as shown in Fig. 22. Now move your fingers towards each other until they come together. The stick still balances. You can repeat the trick, changing the initial position of your fingers, but the result will always be the same, whatever you use—a ruler, a walking stick, a broom, or a billiard cue. The stick balances.

Why?

First of all you must realise that since the stick balances when the fingers come together, they must consequently be right under the stick's centre of gravity (a body balances when the perpendicular from the centre of gravity falls within the supporting base). When the fingers are parted, a greater pressure is exerted on the finger closer to the stick's centre of gravity.

Fig. 22

Ruler experiment.

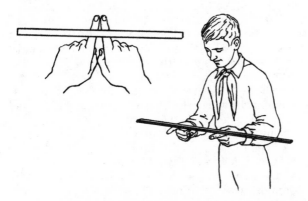

The greater the pressure, the greater the friction. Consequently, the finger closer to the centre of gravity experiences a greater friction than the finger further

Fig. 23

Same experiment but with a broom. Why don't the scales balance?

away. That is why the finger closer to the centre of gravity does not slide beneath the stick. The finger that does slide is the one further away from the centre of gravity. As soon as one finger approaches the centre of gravity the other one begins to slide until, finally, the two fingers come together. As only one of the fingers—the one further from the centre of gravity—moves each time, it is only natural to see them eventually come together right beneath the stick's centre of gravity.

Repeat this experiment with a broom (Fig. 23, top). Then ask yourself: Suppose we cut the broom in half where it balances when the fingers come together, and place each piece on the scale pans (Fig. 23, bottom). Which would be heavier—that with the stick or that with the broom? You might think that since the two parts balanced before on your fingers they ought to balance again on the scales. Actually the pan with the broom is heavier. The answer is simple enough. Realise that when you balanced the broom on your fingers, the forces exerted by the weights of both parts were applied to the unequal arms of a lever. On the scales, though, the same forces are applied to the two ends of an equal-armed lever.

For the "Science for Entertainment" pavilion in Leningrad's Recreation Park I ordered a set of sticks with differently situated centres of gravity. These sticks could be unscrewed into two usually unequal parts, exactly at the centre of gravity. Visitors were surprised to see that when these two pieces were placed on scales the shorter part proved to be heavier than the longer one.

3 Rotation

Why Does Not a Spinning Top Topple over?

Few of the thousands who have spun a top in their childhood will give, I imagine, the right answer. Indeed, why doesn't a gyrating top, be it upright or even inclined, topple over—as one might expect it to do? What is it that keeps it in its seemingly unstable position? After all wouldn't gravity affect it?

Here we have a rather curious interaction of forces. Now since the theory relating to the spinning top is no simple affair, I shan't bother to explain it but merely furnish the main reason.

Fig. 24 depicts a top spinning in the direction designated by the two arrows. Pay heed to segment *A* and its opposite *B*. Segment *A* seeks to move *away from you* while segment *B* seeks to do the opposite. Watch how *A* and *B* move when you tilt the top *towards yourself*. *A* moves upwards and *B* downwards: the two segments receive an impetus at right angles to their own motion. But since in its rapid rotation the circumferential speed of the segments is very great, the insignificant speed that you impart to it will, when added to the great circumferential speed of a point, produce a resultant speed of practically the same value as that of the circumferential speed and thus the top's motion hardly changes. Now you will understand why the top seems to resist attempts to push it over. The greater the top's mass and the

Fig. 24

Why doesn't the spinning top fall?

more rapidly it rotates, the greater the resistance it offers to efforts to topple it.

The explanation is directly linked with the laws of inertia. Every atom in the top moves along a circular orbit in a plane perpendicular to the axis of rotation. In conformity with the laws of inertia this atom or particle seeks at every moment to veer off its circular orbit on to a straight line tangential to the orbit. However, since every tangential path is situated in the same plane as the circumference itself, every particle always tries to contain its motion within the plane perpendicular to the axis of rotation. This means that all planes in the top perpendicular to the axis of rotation try to stay put in their original position in space and that, consequently, the common perpendicular to them—the axis of rotation itself— also seeks to retain its original direction.

I shan't examine all the motions of the top that are produced when an external force is brought to bear, as it would call for a too detailed, and hence, probably, boring explanation. I merely wish to tell you why every rotating body seeks to prevent the direction of its axis of rotation from changing.

This property is widely drawn upon in modern engineering. Navigators and airmen employ sundry gyroscopic devices such as compasses and gyroscopes

Fig. 25

A spinning top hit upwards retains the initial direction of its axis.

that are all based on the gyroscope principle. Rotation imparts stability to shells and bullets and can also be drawn upon to ensure stability in flight in the case of such space vehicles as satellites and rockets. These are the advantages that are to be derived from what one would have thought just a simple toy.

Juggling

Many of the amazing and varied tricks that jugglers perform are also based on the property of rotating bodies to retain the direction of the axis of rotation. Let me now quote from that fascinating book called *Spinning Tops* by Prof. John Perry, a British physicist.

"I once showed some experiments on spinning tops to a coffee-drinking, tobacco-smoking audience in that most excellent institution, the Victoria Music Hall in London.... I impressed my audience as strongly as I could with the above fact, that if one wants to throw a quoit with certainty as to how it will alight, one gives it a spin; if one wants to throw a hoop or a hat to somebody to catch upon a stick, one gives the hoop or hat a spin; the disinclination of a spinning body to let its axis get altered in direction can always

Fig. 26   Fig. 27

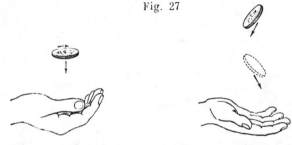

How a spinning coin behaves when flicked up.

How the same coin behaves when it is flicked up without spinning.

be depended upon. I told them that this was why smooth-bore guns cannot be depended upon for accuracy; that the spin which an ordinary bullet took depended greatly on how it chanced to touch the muzzle as it just left the gun, whereas barrels are now rifled, that spiral grooves are now cut inside the barrel of a gun, and excrescences from the bullet or projectile fit into these grooves, so that as it is forced along the barrel of the gun by the explosive force of the powder, it must also spin about its axis. Hence it leaves the gun with a perfectly well-known spinning motion about which there can be no doubt. ... Well, this was all I could do, for I am not skilled in throwing hats or quoits.

But after my address was finished ... two jugglers came upon the stage, and I could not have had better illustrations of the above principle than were given in almost every trick performed by this lady and gentleman. They sent hats, and hoops, and plates, and umbrellas spinning from one to the other. One of them threw a stream of knives into the air, catching

Fig. 28

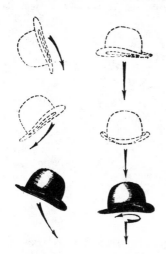

You will catch the hat much more easily if you give it a twirl before throwing it up.

them and throwing them up again with perfect precision, and my now educated audience shouted with delight and showed in other unmistakable ways that they observed the spin which that juggler gave to every knife as it left his hand, so that he might have a perfect knowledge as to how it would come back to him again. It struck me with astonishment at the time that, almost without exception, every juggling trick performed that evening was an illustration of the above principle."

New Solution for Columbus and His Egg

Columbus solved his famous problem of how to stand an egg on its end very simply. He just cracked the top. Incidentally the very popular myth of Columbus and the egg is not really true. Hearsay has ascribed to the celebrated explorer what another man did much earlier and for a totally different reason.

Fig. 29

How to solve Columbus' egg problem. The egg stands on its end when spun.

This was the Italian architect Brunelleschi who built the tremendous dome of the Florentine Cathedral and who claimed that "My dome will keep in place just as reliably as this egg stands on its narrow end."

Actually Columbus' solution is all wrong. When he cracked the egg Columbus changed its *form*, which means that it was already not an egg but a different body that he caused to stand on its end. The crux of the problem lies precisely in the egg-shaped form. As soon as we change the shape, we substitute another body for the egg. The solution that Columbus provided is really not for the body for which it was meant. But

64

we shall be able to solve the great navigator's problem without changing the egg's shape by invoking the property of the top. Just spin the egg on its end. It will stand for a time without falling down on its broad or even narrow end. Fig. 29 shows you how to do it. Be sure to use only a boiled egg. This, incidentally, can't be at variance with the terms of Columbus' problem, because when he propounded it he must have taken an egg straight from the dinner table, and so it must have been a boiled one. You won't get a raw egg to spin on its end, because of its brake-acting liquid contents. This incidentally offers a very simple way—one that many housewives know—of how to tell between a raw and a hard-boiled egg.

Gravity "Destroyed"

"Water will not pour out of a rotating vessel—even when the vessel is bottoms up. Rotation prevents it."

Aristotle wrote that some two thousand years ago. Fig. 30 depicts this effective experiment which no doubt is familiar to many of you. If you swing a pail of water quickly enough—the way shown in the figure—you will be able to prevent the water from spilling out, even when the pail is turned bottoms up.

People usually explain this as due to a "centrifugal force", which they understand as an imagined force supposedly applied to the body and responsible for its tendency to shoot away from the centre of gravity. This "force" is nonexistent; the tendency mentioned is nothing else than a manifestation of *inertia* and every inertia-caused motion needs no force to get it going. Physicists understand the centrifugal force as something entirely different—as the actual force with which a rotating body pulls at a string holding it or presses against its curvilinear orbit. This force

65

is applied not to a moving body but to the *obstacle* that prevents it from moving rectilinearly—a string, the rails of a curved track section, etc.

Back now to our swinging pail. Let us see if we can find the reasons for this phenomenon, without resorting at all to the ambiguous conception of the "centrifugal force". Ask yourself: If I make a hole in the pail, in which direction will the water spurt out? If

Fig. 30

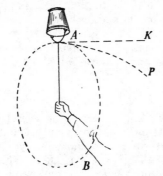

Why doesn't the water spill out?

there were no force of gravity it would—by inertia—spill out at the tangent AK to the circumference AB (Fig. 30). Gravity however will make it bend to describe the parabolic curve AK. If the circumferential speed is large enough, the curve will lie outside the circumference AB, showing the direction the water in the swinging pail would take, were it not obstructed by the pail. That shows you that the water does not at all seek to move vertically downwards, which is why it doesn't spill out. It would spill out only if the top of the pail were to face in the direction of its rotation.

Reckon now the speed with which we must rotate the pail so that the water would not spill down. The centrifugal acceleration of the rotating pail must be no less than the acceleration of gravity. Then the path which the water in the pail will seek to follow will lie outside the circumference described by the pail and at no point will the water lag behind the pail. The formula we must employ to calculate the centrifugal acceleration W is

$$W = \frac{v^2}{R},$$

in which v is the circumferential speed, and R the radius of the circular orbit. Since the surface acceleration of gravity is $g = 9.8$ m/sec^2, we thus get the inequality

$$\frac{v^2}{R} \geqslant 9.8.$$

Presuming that R is equal to 70 cm,

$$\frac{v^2}{0.7} \geqslant 9.8 \text{ and } v \geqslant \sqrt{0.7 \times 9.8}\,; \quad v \geqslant 2.6 \text{ m/sec.}$$

To obtain a circumferential speed of this order we must make about one and a half turns every second. This is quite feasible. Our experiment will cause us no particular difficulty.

Engineers draw on the tendency of liquids to press on the walls of a vessel in which they are rotated around a horizontal axis, for what is called *centrifugal casting*. The essential point here is that a heterogeneous liquid stratifies according to specific gravities, with the heavier and lighter components disposing themselves respectively further away from and closer to the axis of rotation. As a result all the gases that molten metal contains and which produce blowholes, seep out into the casting's central hollow channel. Castings produced by this method have good structure and density and are free from blowholes. The method is cheaper than ordinary pressure die casting and requires no intricate equipment.

You as Galileo

People who like heavy sensations may derive some amusement from a very original attraction called the "devil's swing". There was one in Leningrad once. Since I never tried it, I shall take the liberty of quoting from Fedo's collection of scientific amusements to describe it.

"The swing is suspended to a firm horizontal bar spanning the room at a certain height from the floor. After everybody has taken his seat the attendant locks

Fig. 31

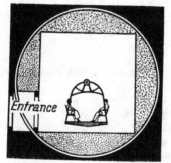

Devil's swing.

the door, removes the board used to mount the swing and, after telling the people on it that he will now take them on a short air trip, gives the swing a push. Then he stands at the back very much like a footman on a carriage, or goes out. Meanwhile the swing goes higher and higher until it finally describes a full circle. The swing goes faster and faster, quite noticeably, and the people on it experience an unquestionable sensation—even though most were forewarned—thinking themselves to be flying through space upside down and involuntarily grip at their seats so as not to fall out. Then the swinging diminishes until the swing finally comes to a complete halt a few seconds later.

"Actually the swing was *stationary all the time*. It was the room itself which with the aid of a very simple mechanism was made to revolve around the people seated on the swing. Pieces of furniture are attached to the floor or walls. The lamp, welded onto the table so as to turn easily, had an electric bulb shielded by a large shade. The *attendant* who seemed to be pushing, actually did no more than to make the swinging accord with the room's slight oscillations and only pretended to be pushing. The entire setting makes for a very successful deception."

As you see the secret is as easy as pie. But I am sure that you succumb to the illusion even though you now know all about it because it is so strong.

Pushkin has a little poem called "Motion":

"There is no motion"—quoth the bearded sage.*
His interlocutor** in answer started walking;
An apt reply—more eloquent than talking
Or mincing words upon the printed page.
However, gentlemen, that most amusing case
Reminds me of another, somewhat trite;
Although the Sun seems moving on clear days,
Yet it was stubborn Galileo who was right.

---

* The Greek philosopher Xenon, who lived circa 500 B.C. and who claimed that everything in the world was stationary. "We thought things were moving," he alleged, "only because of deceptive sensory illusions."

** Diogenes.

Among the people seated on the swing and not initiated into its secrets you would be a Galileo—only in reverse. While Galileo argued that the sun and stars were stationary and that we ourselves were revolving around them contrary to appearances, you would argue that the swing was stationary and that the entire room was revolving around it. Most likely you too would experience Galileo's sad lot, and would be regarded as a decrier of the seemingly self-obvious. ...

Take up the Point with Me

Do you think it will be easy for you to prove your point? If you do, you're mistaken. Suppose you are seated on the "devil's swing" and want to convince your neighbours that they are labouring under a delusion. I suggest you take up the point with me. Before we begin, let us wait until the swing seems to be describing a full circle. There is one condition: you must stay seated on the swing while we argue. We'll take everything we need in advance.

You: How can you doubt that it is the room which is revolving and that we are stationary? After all, if our swing really turned upside down, we would fall out. We certainly couldn't hang in the air upside down. But since we aren't falling out, that means that it is the room which is turning and not the swing.

I: Remember the water in the rotating pail. It didn't spill out, did it, even though the pail was upturned? Or take the cyclist in the "looping the loop" attraction. He doesn't fall either, even though he rides with his head upside down.

You: Let us then figure out the centrifugal acceleration to see whether it is big enough to prevent us from

70

falling out. Since we know how far away we are from the axis of rotation and how many turns we are making each second, we can easily deduce from the formula. ...

I: Don't bother. The attendant told me that we would be making quite enough turns to explain the whole thing as I say. So, whether you figure it out or not, it won't help.

You: But I haven't lost hopes of convincing you. As you see, the water in the glass doesn't spill out. But, I suppose, you will again refer to the rotating pail. Go ahead. I have a plumb-line here, and I see that it points down all the time. If we were rotating and the room were stationary, the plumb-line would swing with us as it would always point down.

I: You are mistaken. If we were rotating fast enough the plumb-line ought to be thrust away from the axis all the time in the direction of the radius of rotation, or, in other words, it should point at our feet. And that is exactly what it does.

How to Clinch the Argument

This is what you should do to convince your opponents. Take a spring balance with you when you seat yourself on the swing, place a weight of one kilogramme, say, on its pan, and watch the pointer. The fact that it will remain constant proves your point.

After all, if we were indeed circling about the axis together with the spring balance, the weight would be affected, apart from the force of gravity, also by the centrifugal force, tending at the bottom to add to and at the top to subtract from, the weight. But since the balance pointer remains constant, the weight neither *increasing* nor *reducing*, consequently, it's the room that is turning and not we ourselves.

One enterprising American started at a public amusement park an extremely interesting and instructive merry-go-round. The people who went into this revolving ball-shaped room experienced sensations which one would come across only in the Land of Nod, or in a fairy-tale world.

First recollect what happens to you when you find yourself standing on the edge of a fast spinning round platform. The rotation tries to send you flying off; and the further away you are from the middle of the whirligig the more of a "push" you get. Now shut your eyes. You seem to be balancing with difficulty on an inclined—not horizontal!—plane. Why? Let us see what forces act on our body in this particular case (Fig. 32). Rotation tries to throw us off while gravity

Fig. 32

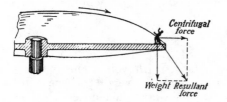

Centrifugal force

Weight *Resultant force*

What one would feel on the edge of a revolving platform.

pulls us down. The two forces combine—according to the rule of the parallelogram—to produce a resultant *inclined downwards*. The faster the spin, the bigger the resultant and the more obtuse the angle.

Imagine now that the edge of the platform is tipped upwards and that you are standing on it (Fig. 33). When the platform isn't moving you won't be able to stand there: you will surely slide down or even fall. But it will be quite different for you when the platform is rotating. Then, provided the platform is spinning quickly enough, this inclined plane will seem horizontal to you because the resultant force of the centrifugal push and gravitational pull will likewise Fig. 33 be inclined, only now at right angles to the tipped

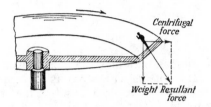

Centrifugal force

You won't fall standing on the inclined edge of a revolving platform.

Weight Resultant force

edge. (This, incidentally, explains why on curved sections of railway track, the outer rail is a bit higher than the inner one. This also explains why racing tracks are steeped inwards and why racers are able to drive along a heavily inclined circular wall.)

If you curve up the edge of the whirligig so that, at a definite speed, *every* surface point of it will be at right angles to the resultant force the person standing on it will think he is on a flat floor, wherever he is. Mathematicians have found this curved surface to be that of a geometrical body called a *paraboloid*. One can produce it by quickly twirling a glass half full of water around a vertical axis. In this case the water at the rim will rise while that in the centre will sink, with the result that its surface will assume the form of a paraboloid. If we were to use some molten wax instead of water and spin the glass until the wax set, we would get an exact paraboloid. When twirled at a definite velocity, the surface of such a solid would present a horizontal plane, and a little ball on it would not roll off but stay put (Fig. 34), wherever placed.

73

Fig. 34

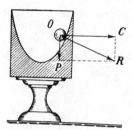

If you spin the tumbler fast enough, the little ball on the side will stay in place.

Now you will easily understand how the "bewitched" sphere is designed. Its bottom (Fig. 35) is a large revolving platform shaped as a paraboloid. Though spun very smoothly by means of a concealed mechanism, the people inside would feel their heads go into a whirl if the furnishings of the "sphere" room were not spinning together with them. To make you think that the platform isn't moving, it is placed inside a large opaque sphere twirling around as fast as the platform itself.

Now what would you see and feel inside it? When it rotates, the floor under your feet would seem flat wherever you stood—on the axis itself where the floor

Fig. 35

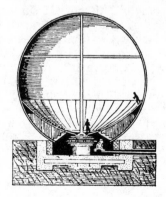

The "bewitched" sphere (cross-section).

74

is really horizontal, or at edge, where it is inclined at 45°. Your eye clearly sees the concavity of the floor but your muscles make you feel that you are standing on an even flat floor. What the two senses register is most contradictory. When you walk from edge to edge of the platform it will seem to you as if the entire sphere had swung over under your weight with the ease of a soap bubble—despite its enormity. While you will think you are standing on a flat floor

Fig. 36

Fig. 37

True position of two men in the "bewitched" sphere.

What each man thinks he sees.

all the time, the other people in this "bewitched" sphere will seem to be crawling up and down the walls like flies (Figs. 36 and 37). Water spilled on the floor would spread out evenly over its concave surface and you would think that you had before you an inclined wall of water.

All your habitual notions of gravity would be gone. An airman making a turn would experience the same. Flying at a speed of 200 km/h along a curved route with a radius of 500 m he would see the ground "loom" and tip at an angle of 16°.

In the German city of Goettingen a similar rotating laboratory was built for scientific purposes. This was

Fig. 38　　　　　　　　　　Fig. 39

Spinning laboratory. True position.

Spinning laboratory. Apparent position.

a cylindrical room (Fig. 38) 3 m across, which did 50 revolutions a second. As its floor was flat the observer standing by its wall thought that the spinning room had leaned backward and that he himself was half lying on an inclined wall (Fig. 39).

Liquid Telescope

The best kind of mirror for a reflecting telescope is a parabolic one, precisely the form the surface of a liquid assumes in a gyrating vessel. Telescope designers take great pains to produce a mirror of this shape, devoting years on end to its manufacture. The American physicist Wood got round the difficulty by devising a liquid mirror. He spun mercury around in a wide-mouthed vessel and obtained an ideal parabolic surface which could well act as a mirror since mercury excellently reflects light. Wood mounted the telescope in a shallow well. The driving belt was used to rotate the vessel containing the mercury and the reflection of Wood's face. The telescope, however, had its drawbacks. The slightest jolt wrinkled the liquid mirror's surface and distorted the image. Despite its tempting simplicity Wood's mercury telescope failed to find practical application. Neither the inventor himself nor contemporary physicists took it seriously. Here, for instance, is a note that A. G. Webster, the head of the physics department of one American university, jotted down after he had seen this original device:

> *Ding, dong, bell,*
> *Prof is in the well,*
> *What did he put in?*
> *A basin full of tin.*
> *What did he get out?*
> *Nothing, just about.*

Looping the Loop

Some of you may have seen the dizzying cycling stunt at the circus. A cyclist loops the loop, riding *upside down* at the top. Fig. 40 gives an idea of this attraction. The performer cycles down the inclined track to glide up swiftly and describe a complete circle, upside down, before reaching ground safely once more. (The stunt was invented in 1902 simultaneously by two circus performers— "Diabolo" Johnson and "Mephisto" Noisette.)

This brain-teasing cycling stunt seems to be the acme of acrobatics. Perplexed, the public wonder what mysterious force keeps the daring performer from falling down and breaking his neck. The sceptics suspect a cunning ruse. But there is really nothing supernatural about it all. Mechanics provides a beautiful explanation. A billiard ball rolling down the same track would perform the trick just as well. At school physics labs you might come across baby "loop-the-loops".

Fig. 40

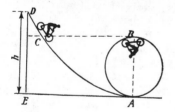

Looping the loop. Bottom left: computation diagram.

To test the stability of the entire arrangement "Mephisto", the cycling celebrity, had used a heavy iron ball of the same weight as he himself plus the bicycle. Only if everything went well did he risk the stunt himself.

You will have guessed by now, of course, that this odd stunt is based on the same principle as that underlying our earlier experiment with the spinning pail of water. To safely shoot the danger zone at the top of the loop the cyclist must work up the adequate speed which depends on the height from which the cyclist takes off. The least allowable speed depends on the radius of the loop itself. This tells you why the stunt doesn't always work out. One must compute accurately enough the height for the take-off, because otherwise the cyclist may break his neck.

Circus Mathematics

I very well know that a string of what would seem dry formulas, "as dry as dry could be," will scare away some lovers of physics. But shunning the mathematical aspect, such people simply rob themselves of the pleasure they might derive by predicting the course of events and determining the conditions required. In the case in question I believe a couple of formulas or so will be quite enough to determine with sufficient accuracy the conditions required to perform successfully such a sensational stunt as the "loop-the-loop".

Here they are. First of all let me designate the values we shall have to deal with. Thus h will be the *height* of the cyclist's take-off, x—part of h above the top point of the loop. Fig. 40 shows us that $x=h-AB$, r—the radius of the loop itself, m—the total *mass* of the performer together with his bicycle (their weight will be expressed as mg, with g being the *acceleration of gravity* which is known to be 9.8 m/sec) and, finally, v the *speed* of the cyclist at the top of the loop.

We can put together two equations from all these values.

(1) Since we know from mechanics that the bicycle's speed at point C of the inclined track, which is at the same level as B (this is the position at the bottom of Fig. 40) is equal to its speed at point B at the top of the loop, our first speed is expressed (we may disregard

79

the energy of the spinning rims of the bicycle wheels as it scarcely affects the result at all) as $v = \sqrt{2\,gx}$ or $v^2 = 2gx$; consequently, the cyclist's speed v at point B is equal to $\sqrt{2\,gx}$, that is $v^2 = 2\,gx$.

(2) To prevent the cyclist from falling on reaching point B centrifugal acceleration must be greater than the acceleration of gravity or, in other words, $\frac{v^2}{r} > g$, $v^2 > gr$ and since we know that $v^2 = 2\,gx$, consequently, $2\,gx > gr$ or $x > \frac{r}{2}$.

So to successfully perform this perplexing stunt the top segment of the inclined section of the track must be higher than the top point of the loop by more than half its radius. The angle at which the track is inclined is not important. The important thing is that the take-off point be more than a quarter of the loop's diameter higher than the highest point of the loop. For a loop of 16 m in diameter the cyclist must start from a height of at least 20 m. If he doesn't he simply won't be able to loop the loop, falling down before he reaches its highest point.

Note that we have disregarded the effect of friction on the bicycle, and consider the speeds at points C and B identical. For that reason we must not make the track too long or sloping, as then, due to friction, the speed the bicycle will have by the time it reaches point B will be less than at point C.

Further note that for this stunt the cyclist does not pedal, letting the machine gather momentum by itself. He cannot and must not accelerate or decelerate his motion. All he must do is to stay in the middle of the track as the slightest deviation may toss him off. The speed with which he loops the loop is great. In the case of a 16-m loop, the cyclist will do the loop in three seconds—the equivalent of a speed of 60 km/h.

It is of course hard to steer a bicycle moving with such a great speed. Actually the cyclist doesn't have to worry about that: he may safely trust the laws of mechanics. "In itself this cycling stunt," a booklet by a professional cyclist tells us, "presents no hazard, provided the arrangement is sturdy and correctly designed. The cyclist himself is the sole hazard. Should his hands shake, should he grow excited and lose his head or should he go giddy all of a sudden, anything might happen."

The famous "Nesterov loop" and other aerobatic tricks are likewise based on the same principle. Of paramount importance in looping the loop is the airman's skilful piloting and adequate starting speed.

"Short Weight"

A certain crank once said that he knew how to "short-weight" customers without cheating them. His idea was to buy goods at the equator and sell them nearer to the poles. It has long been known that at the equator things weigh less than they do at the poles. An article weighing one kilogramme at the equator would weigh five grammes more at the poles. One, however, must use not the ordinary type of scales but a spring balance which would furthermore be made or rather calibrated at the equator. Otherwise one would derive no advantage at all. The article would grow heavier, but so would the weights used in weighing. If we buy a ton of gold in Peru and sell it in Iceland we can make a profit of sorts—provided transportation is free.

Now though I don't think anyone could line his pocket by trading in this fashion, actually our crank was right. Gravity does really increase the further away we get from the equator. This takes place because

81

at the equator due to the earth's rotation, bodies describe the biggest circles and also because at the equator the earth bulges as it were. However the main reason for "short weight" is the earth's rotation. It is this that makes an article weigh at the equator 1/290th less than at the poles.

The difference in weight when transporting a light article from one latitude to another is negligible, but in the case of extremely heavy objects it may be rather impressive. I suppose you never knew that a train weighing sixty tons in Moscow grows 60 kilogrammes heavier when it reaches Arkhangelsk and sixty kilogrammes lighter when it gets to Odessa. There was a time when Spitzbergen shipped as much as 300,000 tons of coal every year to southern ports. Were this amount delivered to some equatorial port, one would have found himself 1,200 tons short—naturally, provided the load were reweighed on delivery by means of a spring balance brought from Spitzbergen. A battleship that weighs 20,000 tons in Arkhangelsk becomes some 80 tons lighter in equatorial waters. We never notice this, however, because everything else, including the water in the ocean too, grows correspondingly lighter. That, incidentally, is why a ship draws the same water at the equator as in the Arctic Ocean: though it becomes lighter, the water it displaces grows lighter by exactly the same amount.

If the earth were to rotate faster than it does now—in other words, if we had a day of not 24 hours but only of 4—the difference in the weight of articles at the equator and at the poles would be much more pronounced. In that case a weight of one kilogramme at the poles would weigh only 875 grammes at the equator. This is roughly the sort of conditions we have on the planet Saturn: near its poles all objects weigh one-sixth more than they do at its equator.

Since centrifugal acceleration increases in direct proportion to the square of velocity we shall easily be able to figure out how fast the earth should rotate at the equator for this centrifugal acceleration to increase by 290 times or, in other words, balance the force of gravity. This would take place if the earth rotated nearly 17 times faster than now (17 times 17 is almost 290). Then bodies would no longer exert any pressure and *would weigh nothing.* On Saturn the same thing would happen, if it rotated but 2.5 times faster than now.

Gravitation

Is the Force of Gravity Significant?

"Were we not to see falling bodies every minute, we would count it a most surprising thing," wrote the famous French astronomer Arago. By force of habit we regard gravity, the attraction the earth has for everything on it, natural and common. But when we are told that objects also attract *each other*, we can hardly believe it, since, generally, we never notice it.

Why is it, indeed, that the law of gravitation does not manifest itself always? Why do we never see tables, water-melons, or people attracting one another? Because for small objects the force of attraction is exceedingly small. Here is a graphic instance. Two persons at a distance of two metres apart do pull at each other. But the force exerted is minute, being under 0.01 mg for people of average weight. In other words, two persons pull at each other with the same force that a 0.00001-gramme weight exerts on the scale pan. Only the extremely sensitive type of scales that scientists use in their laboratories will be able to register such a tiny weight. It goes without saying that this force will never affect us, being completely offset by the friction between our soles and the floor. To push a person standing on a wooden floor—where the friction between his soles and the floor is equivalent to 30 per cent of his weight—you must exert a force of at least 20 kg. And to compare that with the negligible pull of the hundredth of a milligramme is simply ridiculous. A milligramme is a thousandth of a gramme: a gramme is a thousandth of a kilogramme, consequently 0.01 mg is exactly half of a *thousand*

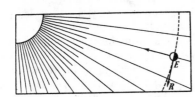

Fig. 41

Sun's pull curves the earth's (*E*) trajectory. Inertia would make the earth strike off at the tangent *ER*.

millionth of the force needed to pull a person away. Small wonder that in ordinary circumstances we never notice the mutual attraction that exists between objects.

But if friction were nonexistent, there would be nothing to prevent even the faintest of pulls from bringing bodies together. In our case of a force of 0.01 mg, however, the *speed* with which the two persons would gravitate towards each other would be negligible. In the absence of friction two persons, 2 metres apart, would draw closer to each other by 3 cm in the first hour, by 9 cm in the next hour, and by 15 cm in the third hour; as you see, speed increases, the closer the two draw together, nevertheless it would take another five hours before the two persons would be drawn together.

Gravity is evident in cases when friction presents no obstacles, or, in other words, in the case of bodies at rest. It will act on a weight hanging on a piece of thread, causing it to dangle vertically down. However should there be some massive object hard by, it will pull the weight towards itself, making the thread deviate slightly from the vertical in the direction of the resultant between gravity and the pull of the massive object. This was first observed by Maskelyne in 1775 near a big mountain in Scotland. He compared the deflection of the plumb-line with the direction towards the celestial pole on either side of the mountain. Subsequent more elaborate experiments conducted with the aid of specially devised scales, enabled scientists accurately to gauge the force of gravity.

The force of attraction between small masses is negligible. As the masses increase, so does the force of attraction increase accordingly, in direct proportion to their product. Many, however, are prone to exaggerate this force. A certain scientist—true not a physicist but a zoologist—tried to assure me that the frequently observed mutual attraction between ships at sea was due to gravitation. A reckoning will show at once that gravitation has no share in it. Two 25,000-ton battleships will attract each other with a force of only 400 g, when they are 100 metres apart. This is naturally far too little to cause the ships to move. We shall explain the real cause of this mysterious affinity between ships in the next chapter but one, dealing with the properties of liquids.

Though negligible in the case of small masses, gravitation proves to be quite a force when we have to do with the colossal masses of the celestial objects. Even the distant planet of Neptune, right out on the rim of the solar system attracts the earth towards itself with a force of 18 million tons. Despite the tremendous distance between us and the sun, it is only gravitation that keeps the earth from straying off its orbit. Were the sun to cease exert its attraction, this planet of ours would dash off at a tangent to its orbit to race headlong into the infinite reaches of space.

Steel Earth-Sun Cable

Imagine for a while that the sun's powerful attraction had indeed ceased to exist and that the earth was faced with the fantastically horrible prospect of being lost in the bleak and murky depths of the universe. Imagine further that a group of engineers had found a way of saving the situation, that they had thought of substituting for the invisible "hawser" of gravitation a sturdy steel cable that would keep the earth on its round-the-sun orbit. After all what could be stronger than steel which can stand a pull of 100 kg to every sq mm? Suppose you have a thick steel column five metres across. Since its cross-section has an area of 20,000,000 sq mm, this column will break only under a weight of 2,000,000 tons. Suppose next that this column would stretch all the way from earth to sun and be clamped down firmly at both ends. Do you know how many of these columns would be needed to hold the earth on its orbit? One million million! To get a clear picture of this dense forest of steel columns studding all the continents and oceans let me add that if they were evenly distributed over that side of the earth turned towards the sun, the space between two neighbouring columns would scarcely be any larger in area than the column itself. When you think of the force that one would need to break this vast jungle of steel columns you will finally realise how mighty is the invisible force that attracts the earth to the sun.

This stupendous force manifests itself by curving the earth's path, swinging it away from a tangent by

3 mm every second. That is the only reason why the earth travels along a locked elliptical orbit. Isn't it amazing to think what a gigantic force is required to shift the earth merely by 3 mm—the height of this line—every second? This just shows you how tremendous the earth's *mass* is—if such a monstrous force is able to shift it only by 3 mm.

Can We Rid Ourselves of the Force of Gravity?

We have just been wondering what would happen were there no mutual attraction between sun and earth. I told you that in that particular case the earth, having shaken off the invisible chains of attraction, would race headlong into the depths of the universe. What would happen to us and everything around us here on our planet, if the force of gravity suddenly disappeared? There would be nothing to pin us down and so the slightest push would send us hurtling out into space. We really wouldn't even need a push, since the earth's rotation would be quite enough to hurl into space every feebly cohering thing.

H. G. Wells chose this as the theme for his novel *The First Men in the Moon*, in which, describing a fantastic journey to the moon, he produced a very original and clever means for interplanetary travel. Wells' main character is a scientist who invents a special alloy with the curious property of being impervious to forces of gravity. When a screen, made of this alloy, was placed between the earth and an object, the planet's pull was rendered null and disappeared and the object was immediately attracted by other bodies. Wells called this alloy Cavorite after the name of its fictitious inventor.

"Now all know substances are 'transparent' to gravitation. You can use screens of various sorts to cut off the light or heat, or electrical influence of the sun, or the warmth of the earth from anything; you can screen things by sheets of metal from Marconi's rays, but nothing will cut off the gravitational attraction of the sun or the gravitational attraction of the earth. Yet why there should be nothing is hard to say. Cavor did not see why such a substance should not exist, and... he believed he might be able to manufacture this possible substance opaque to gravitation....

"Any one with the merest germ of an imagination will understand the extraordinary possibilities of such a substance. ... For example, if one wanted to lift a weight, however enormous, one had only to get a a sheet of this substance beneath it and one might lift it with a straw."

Using this wonderful alloy, Cavor and his friend built a spaceship on which they undertook a daring journey to the moon. Their vehicle was a very simple affair: it had no engine, being propelled solely by the gravitational attraction of the celestial bodies. Wells describes the spaceship as follows:

"Imagine a sphere, large enough to hold two people and their luggage. It will be made of steel, lined with thick glass; it will contain a proper store of solidified air, concentrated food, water distilling apparatus, and so forth, and enamelled, as it were, on the outer steel—Cavorite. The inner glass sphere can be air-tight, and, except for the manhole, continuous, and the steel sphere can be made in sections, each section capable of rolling up after the fashion of a roller blind. These can easily be worked by springs, and released and checked by electricity conveyed by platinum wires fused through the glass. All that is merely a question of detail. So you see, that except for the thickness of the blind rollers, the Cavorite exterior of the sphere will consist of windows or blinds, whichever you like to call them. Well, when all these windows or blinds are shut, no light, no heat, no gravitation, no radiant energy of any sort will get at the inside of the sphere, it will fly on through space in a straight line, as you say. But open a window, imagine one of the windows open. Then at once any heavy body that chances to be in that direction will attract us. ...

"Practically we shall be able to tack about in space just as we wish."

How Cavor and His Friend Flew to the Moon

Wells furnishes a very interesting description of the spaceship's departure. The vehicle's thin outer skin of Cavorite renders it completely weightless and thus able to shoot up to the top of the ocean of air— much like a cork released at the bottom of a lake. It then continues further—don't forget the inertia of the earth's rotation!—on its free flight through space beyond the boundary of the atmosphere. When Cavor and his friend found themselves in space they manipulated shutters and drew on the gravitational attraction now of the sun, now of the earth, and now of the moon, until they finally reached our satellite. Later on one of the travellers returned to earth in the same projectile.

In this book I shall not bother to analyse Wells' project. Let us, on the contrary, take Wells' story at its face value for a time and go with Cavor and his friend to the moon.

Lunar Half-Hour

What did they experience in a world with a force of gravity that is much weaker than the earth's? Read the story as told by one of the earthmen. They have just landed on the moon.

"I went on unscrewing. ... I knelt, and then seated myself at the edge of the manhole, peering over it. Beneath, within a yard of my face, lay the untrodden snow of the moon. ... Cavor... stretched out his hand for his blanket, thrust his head through its central hole, and wrapped it about him. He sat down on the edge of the manhole, he let his feet drop until they were within six inches of the lunar ground. He hesitated for a moment, then thrust himself forwards ... and stood upon the untrodden soil of the moon.

"As he stepped forward he was refracted grotesquely by the edge of the glass. He stood for a moment looking this way and that. Then he drew himself together and leaped.

"The glass distorted everything, but it seemed to me even then to be an extremely big leap. ... He seemed twenty or thirty feet off. He was standing high upon a rocky mass and gesticulating back to me. Perhaps he was shouting—but the sound did not reach me. But how the deuce had he done this?...

"In a puzzled state of mind, I too dropped through the manhole. I stood up. Just in front of me the snow-drift had fallen away and made a sort of ditch. I made a step and jumped.

"I found myself flying through the air, saw the rock on which he stood coming to meet me, clutched it and clung in a state of infinite amazement. ... Cavor bent down and shouted in piping tones for me to be careful. I had forgotten that on the moon ... my weight was barely a sixth what it was on earth. But now that fact insisted on being remembered. ...

"With a guarded effort I raised myself to the top, and moving as cautiously as a rheumatic patient, stood up beside him under the blaze of the sun. The sphere lay behind us on its dwindling snow-drift thirty feet away. ...

"'Look!' said I, turning, and behold, Cavor had vanished.

"For an instant I stood transfixed. Then I made a hasty step to look over the verge of the rock. But in my surprise at his disappearance I forgot once more that we were on the moon. The thrust of my foot that I made in striding would have carried me a yard on earth; on the moon it carried me six—a good five yards over the edge. For the moment the thing had something of the effect of those nightmares when one falls and falls. For while one falls sixteen feet in the first second of a fall on earth, on the moon one falls two, and with only a sixth of one's weight. I fell, or rather I jumped down, about ten yards I supposed. It seemed to take quite a long time, five or six seconds, I should think. I floated through the air and fell like a feather, knee-deep in a snow-drift in the bottom of a gully of blue-grey, white-veined rock.

"I looked about me. 'Cavor!' I cried, but no Cavor was visible.

"'Cavor!' I cried louder. ...

"Then I saw him. He was laughing and gesticulating to attract my attention. He was on a bare patch of rock twenty or thirty yards away. I could not hear his voice, but 'Jump!' said his gestures. I hesitated, the distance seemed enormous. Yet I reflected that surely I must be able to clear a greater distance than Cavor.

"I made a step back, gathered myself together, and leaped with all my might, I seemed to shoot right up in the air as if I should never come down. ...

"It was horrible and delightful, and as wild as a nightmare, to go flying off in this fashion. I realised my leap had been altogether too violent. I blew clean over Cavor's head."

The following episode from *On the Moon*, a novel by the celebrated Soviet inventor K. E. Tsiolkovsky, will tell us how the force of gravity affects motion. Because of the earth's atmosphere—which impedes the motion of all bodies in it—we fail to see the simple laws of falling bodies; they are complicated by sundry additional factors. The moon, however, hasn't got any atmosphere, and would make a fine laboratory to study falling bodies—provided we could get to it and do scientific research there.

Now let me give the floor to Tsiolkovsky—with the explanation that the two persons talking are on the moon and at the moment wish to see what the flight of a bullet fired from a gun would be like there.

"'But will gunpowder do its job here?'

"'In a void the effect of explosives should be even greater than in air, as the latter prevents them from expanding. As for oxygen, they contain as much as they need.'

"'Let's aim the rifle vertically to make it easier for us to find the bullet later.'

"There was a flash and a slight plop (the kind of sound one would hear transmitted through the ground and human bodies and not through the air of which the moon has none) and the ground quivered.

"'Where's the wad? It ought to be hard by.'

"'The wad has gone up with the bullet and will most likely keep up with it. Back on the earth the atmosphere prevents the wad from winging after the lead bullet. Here on the moon a feather will fall and fly up in the same manner as a stone. Suppose you pluck a feather from your pillow while I take a little iron ball. You will be able to throw your feather and hit a thing that is even far away, just as easily as I will with my little ball—which, considering its small weight, I'll be able to throw as far as some 400 metres. You'll be able to throw your feather just as far. True, you won't kill anyone with it. You won't even feel it when you throw it. So let us throw projectiles with all our might—I think we're about the same in that respect—at one target—that red chunk of granite, for instance. ...'

"The feather slightly outflew the iron ball, as if drawn on by a strong whirlwind.

"'What can have happened? It's already three minutes since we fired the shot but the bullet still hasn't come back.'

"'Have patience for a couple of minutes. You'll probably see it coming back then.'

"And indeed, a couple of minutes later we felt the ground quiver and saw the wad bouncing near us.

"'The bullet certainly flew quite a time. How high did it get?'

"'Some 70 kilometres up—all because of the small gravitational pull and the absence of air drag.'"

Suppose we check this statement. If we take the rather modest figure of 500 m/sec for the bullet's muzzle velocity—with modern rifles it is one and a half times more—it would rise above the earth, *in the absence of an atmosphere*, to

$$h = \frac{v^2}{2\,g} = \frac{500^2}{2 \times 10} = 12{,}500 \text{ m}$$

or 12.5 km. However since the moon's gravitational pull is but a sixth of the earth's, g will be six times smaller and consequently the bullet will go up to 12.5×6=75 km.

So far we know very little as to what takes place deep down inside the earth's bowels. Some think a molten mass lies beneath the solid crust at the top—which is supposed to be a hundred kilometres thick. Others believe the earth to be solid all the way through. It is hard to say who is right. The deepest well drilled has not been more than 7.5 km down, while

Fig. 42

Drilling a hole through the globe along its diameter.

the deepest mine sunk and down which man has gone is only 3.3 km down*.

The earth's radius meanwhile is 6,400 km. We would be able to say for sure if we could drill a hole right through the earth along its diameter. Unfortunately our present well-drilling equipment can't do that—even though all the wells drilled so far add up to more than the earth's diameter in length.

In the 18th century the mathematician Maupertuis and the philosopher Voltaire dreamed of tunnelling right through the earth, and later the French astronomer Flammarion brought up the project again—true on a more modest scale. Fig. 42 reproduces the drawing he gave as a preface to his article about it.

---

* A gold mine in Boxburg, the Transvaal, South Africa, which moreover has its mouth 1,600 m above sea level, meaning that it is only 1,700 m down from sea level.—Ed.

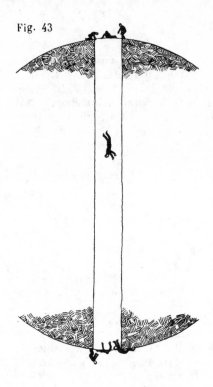

Fig. 43

Should you ever chance to fall into a shaft drilled through the earth's centre you would bob endlessly pendulum-like from end to end. It would take you 84 minutes to get from end to end.

Of course no tunnel of this nature has been dug yet; however we shall suppose it has been dug, in order to deal with a curious problem. What do you think would happen if you fell into such a bottomless well (discount air drag for the time being)? You won't crash into the bottom, because there isn't one. But where would you stop? It wouldn't be in the centre of the earth because by the time you reached that point your velocity would be so great—about 8 km/sec— that you would simply overshoot and continue to fall, gradually slowing down, until you reached the outlet at the opposite end. Here you must grab the edge for dear life, because otherwise you would again shoot back to the other end. If you again failed to grab the edge, you would simply swing on endlessly, like a pendulum—of course, provided air drag is disregarded (if we did take it into account, oscillation would gradually fade and you would finally wind up in the centre of the earth).

Now how long would it take you to bob from end to end and back? The answer is: 84 minutes 24 seconds, or roughly an hour and a half.

"That would be so," Flammarion continues, "if our well were dug from pole to pole along the axis. As soon as we shift the opening to any other latitude in Europe, Asia or Africa, we must take the earth's rotation into account. We know that every point of the earth does 465 m/sec at the equator and 300 m/sec at the latitude of Paris. Since circumferential velocity *increases* the further away we are from the axis of rotation, a plumb-line dropped in a well will deflect eastward from the vertical. If we dug our bottomless well at the equator, we should have to make it either very wide or greatly inclined, because a body dropped into it from the surface would fly way east of the centre of the earth.

"Further if the mouth were on a South-American plateau two kilometres above sea level, while the outlet were at ocean level, anybody inadvertently falling into the opening would fly out at the other end two kilometres up. However if both openings were at ocean level, you could grab the person by the hand when he appears at the mouth, because here his speed would be nothing. In the previous case you would better beware of the 'traveller hurtling by'."

Fairy-Tale Railway

A booklet with the odd title of *A Motorless Underground, St. Petersburg-Moscow Railway, Fantasy in*

97

Three as Yet Unfinished Chapters was once put out in St. Petersburg (now Leningrad). Its author, A. A. Rodnykh, suggested a very witty project which those fond of physical paradoxes would find not devoid of interest. His idea was to "dig a 600-km-long tunnel, linking up the two capitals by an absolutely straight underground line. This would give us for the first time the opportunity to travel along a straight line in-

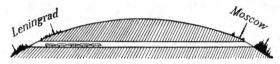

Fig. 44

In this Leningrad-Moscow tunnel trains would need no locomotive. Their own weight would impel them to speed to and fro.

stead of following curved paths as now." (He wishes to say that all our roads describe arcs, as they follow the curve of the earth's surface, while the suggested tunnel would follow a straight line, along a chord.)

This project—if ever realised—would possess a unique characteristic. In this tunnel a train would *move by itself*. Recall the well that struck right through the earth. This Leningrad-Moscow tunnel would be a somewhat similar affair, the difference being that it would follow not the diameter but a chord. True, a cursory glance at Fig. 44 might suggest that since the tunnel is horizontal, there is no reason for gravity to propel the train. That is simply an optical illusion: as soon as you mentally trace radii to the two ends of the tunnel—the radius will show the perpendicular —you will realise that the tunnel is not at all at right

98

angles to them, or in other words that it is not horizontal but inclined.

In such a slanting tunnel every object should swing, due to gravity, to and fro like a pendulum, hugging to the bottom. Inside it a train would move by itself along the rails, its weight doing the work of a locomotive. At first the train will move very slowly, but with every new second its speed would increase, to reach soon a figure so incredible, that the air in the tunnel would offer noticeable resistance. However, let us forget for a while this annoying circumstance which has blocked the realisation of so many fascinating projects and see what will happen to our train. Its speed will be so great—many times that of a cannon ball—when it reaches the middle of the tunnel that it will overshoot it and almost reach the other end. I say "almost" because of friction; were there no friction our train would go all the way from Leningrad to Moscow by itself without a locomotive. It has been reckoned that it would take the train to get from end to end the same amount of time as it took the person who fell into the pole-to-pole tunnel—42 minutes 12 seconds, a period of time which curiously enough does not depend on how long the tunnel is. It would take us the same 42 minutes 12 seconds to travel along similar tunnels between Moscow and Leningrad, between Moscow and Vladivostok or between Moscow and Melbourne, for that matter. (There is one more curious point about the bottomless well—the duration of the swing depends not on the *size* of the planet, but on its *density* alone.)

In this tunnel any other conveyance—be it a motorcar, a cart or anything else—would behave in the same manner. Indeed a fairy-tale road. Itself stationary, it is able to make anything on wheels race along it from end to end and moreover with unimaginable speed!

99

How to Dig Tunnels

Fig. 45 shows three ways of digging a tunnel. Now tell me which of the three is the horizontal one? It is neither the upper nor lower; it is the one in the middle that traces an arc. At every point this middle tunnel will be at right angles to the vertical—or to the earth's

Fig. 45

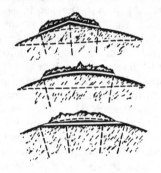

Three ways of tunnelling through mountains.

radius. It is horizontal as its curve fully accords with the curve of the earth's surface.

Long tunnels are usually dug as shown at the top of Fig. 45—along straight lines which swing out at a tangent to the earth's surface from the two ends. This tunnel first goes *upwards* and then slants *downwards*. It is very convenient since water does not collect in it, but flows by gravity towards the ends.

A tunnel dug strictly horizontally would be shaped as an arc. Water would not flow out of it, as it would be in a state of equilibrium at every point. When such a tunnel is more than 15 km long—the Simplon tunnel, for instance, is 20 km long—a person standing at one end would not see a person standing at the other. He would see nothing but the ceiling of the tunnel as its central point would be 4 metres higher than its two terminals.

Finally, if we were to dig the tunnel as a straight line between the two terminals, it would gradually slope down towards the middle. In such a tunnel, far from flowing out, water would, on the contrary, gather in the middle, which would be the lowest part. On the other hand a person standing at one end would see his friend at the other end—as you may deduce from the figure. (Incidentally, while all horizontal lines *curve*, and can by no means be straight, vertical lines, on the contrary, can only be straight.)

Travelling
in a Projectile

Before winding up our discussions of the laws of motion and gravitational attraction, let us deal with that fantastic journey to the moon so fascinatingly described by Jules Verne in his *From Earth to Moon* and *Around the Moon*. If you have read these books you will most likely remember that after the close of the Civil War the members of the Baltimore Cannon Club decided out of enforced idleness to cast a huge cannon and to shoot at the moon an enormous hollow projectile with passengers inside. Could this ever be done? And firstly, could we ever impart to bodies a velocity great enough for them to escape from the earth never to return?

Newton's Mountain

Newton, the genius who discovered the universal law of gravitational attraction, wrote in his *Principia*, that due to gravity any stone we throw up into the air veers away from a straight path and falls back on the earth, describing a curvilinear trajectory. However, if we impart a greater velocity to the stone, it flies much further and it may very well happen that it might describe an arc of some ten, one hundred, or one thousand miles and finally break away from the earth's fetters never to return. Suppose *AFB* in Fig. 46 is the earth's surface, *C* its centre, while *UD*, *UE*,

Fig. 46

How stones would fall when ejected horizontally and with great speed from the top of a mountain.

UF and *UG* are the curved trajectories described by an object thrown horizontally from a very tall mountain with increasing velocities. (We shall disregard air resistance.) In the case of a smaller initial velocity, the object describes the curve *UD*, in the case of a higher velocity, the curve *UE*, and in the case of still greater velocities, the curves *UF* and *UG*. Provided the initial velocity is adequate, our object will circumvent the earth and return to its starting point at the top of the mountain, and since its velocity will now be *identical* with the initial velocity, it will continue along the same orbit.

If we had a gun mounted on this mountain, the cannon balls it fired would—provided the starting velocity was high enough—never fall back on earth again but girdle the globe endlessly. It is easy enough to figure out (for which see the second chapter of *Physics for Entertainment*, Book One) that this would happen when we have a muzzle velocity of about 8 km/sec. A projectile fired at this speed would orbit the earth as a satellite to race along 17 times faster than any point on the equator and have a period of revolution of 1 hour 24 minutes. When the muzzle velocity is greater, the missile no longer travels round the earth on a circular orbit; it follows a more or less elongated elliptical orbit, shooting out far away from the earth. When the initial velocity is still greater, about 11 km/sec, the missile will escape forever into space. (Note that we have been speaking of the motion of projectiles in a *void*, not in air.)

Let us now see whether we could fly to the moon as Jules Verne suggests. Modern guns impart a muzzle velocity of not more than 2 km/sec, which is only a fifth of the speed we require to fly out to the moon. The Baltimore cannoneers thought, however, that if they cast a giant cannon and used a tremendous charge, they would be able to obtain a high enough velocity to shoot their projectile to the moon.

Fantastic Gun

So the members of the Baltimore Cannon Club had cast an enormous gun a quarter of a kilometre long, which was vertically bedded in the ground. Then a huge 8-ton projectile with a passenger cabin was made. The charge consisted of 160 tons of nitro-cotton. When the gun was fired, the projectile received, according to Jules Verne, a starting velocity of 16 km/sec, which air drag reduced to 11 km/sec—quite enough to fly on to the moon after quitting the earth's atmosphere. That is what Jules Verne thought.

Let us now see whether physics bears him out.

Jules Verne's project is vulnerable but not at the point where readers usually think it to be. In the first place one can prove that gun powder will never be able to impart to artillery projectiles a velocity of more than 3 km/sec.

Furthermore, Jules Verne ignored air drag which, considering the tremendous velocities involved, should be pretty great. At any rate, it would change the trajectory completely. However, in addition to all that there are other serious objections to a moon flight in a cannon ball. It is what lies in store for the passengers themselves, that presents the main stumbling-block.

Don't think the actual journey is dangerous. As long as they survive the firing moment, they will have nothing more to worry about. The colossal speed with which they would be hurtling through space in their projectile would be just as harmless for them as for us earthmen is the still greater speed with which our planet races around the sun.

Heavy Hat

The moment of greatest peril is the few hundredths of a second when the projectile is accelerating in the gun's bore, because in this negligible interval, the velocity is supposed to increase from zero to 16 km/sec. No wonder Jules Verne's cannoneers looked forward to that moment with such trepidation. Barbicane was quite right when he said that the very first moment of flight would be just as dangerous for the passengers inside were they not inside but in front of the missile. Indeed, when the gun is fired, the cabin floor should strike at the passengers from below with the same force as a projectile would strike at anything in its path

Jules Verne's cannoneers made light of this peril thinking that at worst they would get off with a rush of blood to the head. ...

Actually it's much more serious than that. In the barrel the projectile accelerates, its velocity increases by the constant pressure of the gases that the gunpowder gives off when ignited. In the negligible fraction of a second this speed mounts, as I have already remarked, from zero to 16 km/sec. Let us suppose for simplicity's sake that acceleration is uniform. In that case the acceleration required to increase the projectile's velocity to 16 km/sec in so short a time would be in the vicinity of 600 km/sec a second (we shall figure it out later).

This figure is fatal, as you will well realise when I tell you that the usual acceleration of surface gravity is only 10 m/sec^2. (Let me add that the acceleration of a racing car from starting point is not more than 2-3 m/sec^2, while the acceleration of a train smoothly pulling out is only 1 m/sec^2.) It hence follows that at the firing moment everything inside the cabin would press down with 60,000 times its true weight. This means that the passengers would grow several dozen thousand times heavier, which would crush them to pulp at once. Mr. Barbicane's top-hat alone would have weighed at least 15 tons when the gun was fired —more than enough to crush its owner.

True, Jules Verne describes several measures taken to soften the impact. The projectile was fitted out with spring guards and a double water-filled floor. This stretches the impact interval and hence the rapidity with which the speed increases is reduced. But considering the tremendous forces that we have to deal with in this case the advantage thus derived is insignificant. Only a fraction would be sliced off the force pinning the passengers to the floor. I personally don't think it makes the slightest difference whether the hat would weigh 15 or 14 tons; it would steam-roller you in either case.

How to Soften Impact

Mechanics tells us how we can reduce the fatal increase of velocity. We must make the cannon *many times longer*—very much longer if we want the force of "artificial gravity" inside the projectile to equal earth gravity at the firing moment. According to a rough-and-ready reckoning we would have to make the gun at least 6,000 km long. This means that Jules Verne's *Columbiad* would have reached the earth's centre. Only then would the passengers have experienced no particularly unpleasant sensations. They would have merely felt twice as heavy, since the slow increase in speed would have added an apparent weight of the same magnitude to their own.

Incidentally over a short interval of time the human organism can stand up to a several-fold increase in weight without any harmful effect. When we abruptly change our course, out roller-coasting, our weight noticeably increases during these short periods; our body presses down harder on the sled. We can safely bear a threefold increase in weight. Supposing that we could safely stand up even to a tenfold increase in weight over a short period of time, it would be enough to cast a gun "only" 600 km long—which, even then, wouldn't help since it is technically impossible to make such a gun.

Only provided we were able to abide by all these conditions, we could think of realising Jules Verne's project. (We dealt in Book One of *Physics for Entertainment* with a serious omission which the French science-fiction novelist was guilty of when he ventured to describe the conditions of life inside the projectile after it had been shot out and was flying to the moon. He forgot the state of weightlessness that would set in when every object inside the projectile would likewise weigh nothing, as the force of gravity would impart an identical acceleration both to the projectile and every object inside it—see further "The Chapter Jules Verne Did Not Write".)

For All Who Like Mathematics

Some of you would most likely wish to check up on the figures given above. Here are the reckonings—which, true, are only approximate as they are based on the supposition that in the gun barrel the projectile moved with a uniform acceleration (actually acceleration is not uniform).

We shall need the following two equations for uniformly accelerated motion:

velocity v after the t-th second is equal to at, where a is the acceleration. In other words $v = at$;

the path S covered in t seconds is determined by the equation

$$S = \frac{at^2}{2}.$$

We shall now proceed to determine the acceleration of the projectile in the *Columbiad*'s barrel. From the novel we know its length—210 metres, i.e., S, the path that the projectile covers. We also know the ultimate velocity: $v = 16,000$ m/sec. We are now able to determine t, the time the projectile moved in the barrel—presuming, naturally, that we are dealing with a uniformly accelerated motion. So,

$$v = at = 16,000, \quad 210 = S = \frac{at \cdot t}{2} = \frac{16,000\,t}{2} = 8,000\,t,$$

whence

$$t = \frac{210}{8,000} \approx 1/40 \text{ sec.}$$

We thus find that it took only $^1/_{40}$ of a second for the projectile to travel the length of the barrel.

Substituting $t = {}^1/_{40}$ in the equation $v = at$, we get:

$$16,000 = {}^1/_{40}\,a, \text{ whence } a = 640,000 \text{ m/sec}^2.$$

As we see the acceleration of the projectile inside the cannon was 640,000 m/sec², or 64,000 times more than the acceleration of gravity. How long should the gun be then, for the projectile's acceleration to be only ten times more than the acceleration of gravity, to wit: 100 m/sec²?

This is a problem which must be solved in reverse order to the one we just solved. We know that a is 100 m/sec², and that v is 11,000 m/sec (which will be quite sufficient in the absence of air drag).

From the equation $v=at$ we get $11,000=100\,t$, whence $t=110$ sec.

From the equation $S = \frac{at^2}{2} = \frac{at \cdot t}{2}$ we find that the gun should be $\frac{11,000 \times 110}{2} = 605,000$ m, or 605 km long.

In this way do we deduce the figures that shatter the fascinating project of Jules Verne's cannoneers.*

---

* Everything in this chapter is unquestionably right. As for the practical aspect of space flight you will have probably read about it elsewhere.—*Ed.*

Properties of Liquids and Gases

Sea in Which One Never Sinks

There is such a sea in a country with a very ancient history. This, of course, is the famous Dead Sea in Palestine. Its water is so salty that nothing can live in it. Due to the local scorching rainless climate the surface water evaporates. Note, though, that it is only water as such which evaporates. The salt dissolved in it remains making the water still saltier. This explains why the Dead Sea has a salt content not of two or three per cent (by weight) as most seas and oceans but of 27 per cent and even more—the salt content increases with depth.

Thus a quarter of the Dead Sea is made up of the salt dissolved in its water. This sea has been estimated to have a total of 40 million tons of salt.

The water of the Dead Sea exhibits a very curious property precisely because of its saltiness. Since it is much heavier than ordinary sea water, you will never sink in it because your body is much lighter.

We weigh noticeably less than an equal volume of very salty water. Hence, according to the law of buoyancy we would never drown in the Dead Sea; we would pop up to the surface just like an ordinary egg in salt water—which, incidentally, sinks in fresh water.

Mark Twain, the famous American humorist, visited the Dead Sea, and in one of his books he wittily describes the unusual sensations that he and his companions experienced when they bathed in it.

"It was a funny bath. We could not sink, one could stretch himself at full length on his back, with his arms on his breast, and all of his body above a line drawn from the corner of his jaw past the middle of his side, the middle of his leg and through his anklebone, would remain out of water. He could lift his head clear out if he chose.... You can lie comfortably on your back, with your head out, and your legs out from your knees down ... you can sit, with your knees drawn up to your chin and your arms clasped around them, but you are bound to turn over presently, because you are top-heavy in that position. You can stand up straight in water that is over your head, and from the middle of your breast upward you will not be wet. But you cannot remain so. The water will soon float your feet to the surface. You cannot swim on your back and make the progress of any consequences, because your feet stick away above the surface, and there is nothing to propel yourself with but your heels. If you swim on your face, you kick up the water like a sternwheel boat. You make no headway. A horse is so top-heavy that he can neither swim nor stand up in the Dead Sea. He turns over on his side at once."

Fig. 47 shows a very pleasant way of whiling away the time on the surface of the Dead Sea. Because of the specific weight of its water the person shown is able to read a book in the shade of an umbrella warding off the blazing sunshine. The water of the Kara Bo-

Fig. 47

Swimming in the Dead Sea.
(From a photograph.)

gaz Gol, a gulf in the Caspian, and of Lake Elton — with its 27-per-cent salt content—exhibits the same unusual properties. (Incidentally, the specific density of the water in the Kara Bogaz Gol is 1.18. "In this dense water one may swim effortlessly and will never sink however hard one may try to violate Archimedes' principle," the explorer Pelsh has noted in this connection.)

Patients taking salt-water baths experience something very similar to what we have just described. When the water is very salty—as at the Staraya Russa spa, for instance—the patient has to make quite an effort to keep submerged. I heard one lady patient at Staraya Russa complain with disgust that the water "was pushing her out of her bath" and she seemed to think that the management was to blame.

The salt content of the water in different seas va-

ries, due to which ships do not draw identically everywhere. Some of you have probably seen the so-
called "Lloyd mark" near the water-line on a ship's
hull, which designates submergence limits in water of
different densities. For instance, the load mark in
Fig. 48 stands for the limit of submergence, which is:

Fig. 48

Cargo marks on a ship's
water-line. Top right: same
marks enlarged. The letters
are explained in the text.

FW in fresh water,
IS in the Indian Ocean in summer time,
S in salt water in summer,
W in salt water in winter,
WNA in the North Atlantic in winter.

Russia introduced these marks as obligatory in
1909.

Let me note in conclusion that there is a variety
of water, which even in its pure state with no admixtu-
res is much heavier than ordinary water. Its specific
density is 1.1, or 10% more than of ordinary water. In a swimming pool filled with this water even
the novice would never drown. Called *heavy* water its
chemical formula is D_2O (its hydrogen component is
made of atoms twice as heavy as those of ordinary hydrogen and designated by the letter D). There is an
insignificant quantity of "heavy" water in ordinary
water—about eight grammes to every pail.

113

The "heavy" water of the D_2O type—there are seventeen possible types altogether—is now obtained in almost its pure form with but a 0.05% admixture of ordinary water. "Heavy" water is extensively used in nuclear technology and, in particular, in atomic reactors. It is obtained commercially and in large quantities from ordinary water.

How an Icebreaker Works

Do the following experiment when you take a bath. Before getting out pull the plug out but stay prone for a while. As more and more of your body emerges from the water, you feel it grow heavier and heavier. This is a most graphic illustration showing how you regain weight lost in water—remember how much lighter you felt in the bath-tub when it was full. When a whale involuntarily engages in such experiments—being stranded in shallow waters during an outgoing tide—the consequences are fatal. The whale is crushed to death by its own monstrous weight. No wonder whales can live only in water. Buovancy saves from gravity's lethal effect.

Now what can all this have to do with icebreakers, you might wonder? The job an icebreaker does is based on the same physical principle. Since that part of the ship jutting out of the water is no longer offset by the water's buoying force, it acquires its "dry" weight. Don't think the icebreaker cuts through ice merely by exerting an endless pressure with its bow. An ordinary ice boat does, but only when the ice is not very thick.

Real icebreakers, like *Krasin* and *Yermak* and the atomic-powered *Lenin*, work quite differently. The icebreaker brings its bow to bear on the surface of the ice—for which purpose the underwater part of the

114

bow is greatly slanted. As it climbs out of the water, the bow "acquires" its full weight—as much as 800 tons in the case of the *Yermak*—and thus crushes through the ice. To increase pressure, water is often pumped into tanks in the icebreaker's bow.

This method is used when the ice is not too thick. Thicker ice is forced to yield by *ramming*. The icebreaker backs away and then goes full steam ahead, bang into the barrier of ice. Now it is not the ship's weight but the kinetic energy of its motion that is brought into play. The ship is actually used as a kind of battering ram. The blows are so powerful that even solid walls of ice, several metres high, crumble. Here is how the polar explorer N. Markov, who took part in the famous 1932 expedition aboard the *Sibiryakov*, describes the way the ship broke through the ice:

"It was among hundreds of solid ice-packs in the midst of huge icefields that the *Sibiryakov* began its 52-hour battle. For 13 full watches, the engine telegraph rang from 'full steam reverse' to 'full steam ahead', as the vessel rammed the ice, crushed it with its bow, climbed out to break it down and then backed away to strike again. The ice, three-quarters of a metre thick, gave way with difficulty. Each new blow took us only a third of the ship's length."

Where to Look for Foundered Ships

Even many seamen think that ships wrecked at sea never sink to the bottom but hang suspended at a certain depth, where the water supposedly "reaches the appropriate density due to the pressure of upper layers".

The author of *Twenty Thousand Leagues Under the Sea* also subscribed to this view. In one place Jules

115

Verne describes a wreck suspended immobile in the depths. In another chapter he reminds us of ships "rotting as they hang freely suspended in the water".

Is this right?

One might think there is some reason for such statements, since the pressure that the water exerts deep down in the ocean is indeed tremendous. At ten metres down, the water exerts a pressure of 1 kg to every sq cm of the submerged body. At 20 metres down, this is already 2 kg, at 100 metres—10 kg, and at 1,000 metres—100 kg.

We know that in many places the ocean bed lies several kilometres deep, reaching to more than 11 km down at the deepest spots—the Marianne depression in the Pacific. You will easily realise what colossal pressures both the water and everything in it should be subjected to at such tremendous depths.

If we push a corked but empty bottle down to a great depth and then pull it out again, we will find it full of water with the cork inside—all because of the pressure that the water exerts deep down. In his *The Ocean*, the celebrated oceanographer John Murray describes the following experiment. Three glass tubes of different sizes, sealed at both ends, were wrapped in a cloth and placed in a copper cylinder which had holes in it to let the water through. The cylinder was sent down to 5 km and then pulled out. When the cloth was unwrapped, a snow-like mass of crushed glass was found. Pieces of wood sent down to similar depths sunk like bricks later, so heavily compressed were they.

Hence it seems only natural to expect that this terrific pressure should make the water at great depths so dense that even heavy objects would not sink any further—in the same way as an iron weight does not sink in mercury. This is a totally erroneous notion. Experiments have shown that water, like all liquids in general, yields very little to compression. Under pressure of one kilogramme to every square centimetre it will compress only by a twenty-two-thousandth of its volume, and the rate of compression increases by the same degree with every extra kilogramme of pressure. We must make water eight times denser than it is for iron to float in it. But to make it just twice as dense, or, in other words, compress it to half its present volume, one must exert a pressure of 11,000 kg/cm². Provided that in general were possible, we would get a pressure of that order only at a depth of 110 km.

Thus it is clear that any noticeable compression at great ocean depths is out of the question, because even at the deepest spot water loses only five per cent of its volume due to compression. (The British physicist Tate has reckoned that if gravity were suddenly to cease and water become weightless, the level of the water in the oceans would rise by an average of 35 metres as the gravity-compressed water would regain its normal volume. Berger has noted that in this case "the ocean would flood 5 million sq km of dry land which is dry land only because the water in the oceans is compressed".) This would scarcely have any effect on buoyancy—all the more so since all solid objects at these depths are subjected to the same pressure and are consequently compressed too.

There need be no doubt, therefore, that wrecked ships sink to the ocean bottom. "Anything that will sink to the bottom of a tumbler of water," Murray says, "will practically sink to the bottom of the deepest ocean."

I have heard the following objection. If you carefully immerse a glass *bottoms up*, it may stay thus, as it will displace an amount of water weighing as much as the glass itself weighs. A heavier metal tumbler might stay in this position, even below water level, without sinking to the bottom. So it is claimed that a capsized cruiser or any other ship might also stop halfway down. If the air in the ship's compartments has no escape, the ship may sink to a certain depth and stay there. After all quite a few ships sink with their keels pointing upwards. Couldn't it be possible that some of them might have not reached the bottom and are still suspended in the murky ocean depths? And though the slightest push would be enough to disturb their equilibrium, righten them, fill them with water and send them to the bottom, could one expect jolts in the ocean depths—that domain of eternal silence and tranquility, where even the worst of storms has no repercussions?

All these arguments are based on a physical error. An overturned glass *does not submerge by itself*. It must be made to do so *by some external force*—in the same way as a piece of wood or an empty corked bottle. So will an overturned ship remain afloat—never to find itself halfway between top and bottom.

How the Dreams of Jules Verne and H. G. Wells Came True

Present-day submarines even beat Jules Verne's fantastic *Nautilus* on some points. True, their speed is half as much, only 24 knots instead of Jules Verne's 50, and the longest voyage they can make is a round-the-world one, whereas Captain Nemo did twice as much. On the other hand, the *Nautilus* had a displacement of only 1,500 tons, a crew of some thirty men and was able to lie submerged for no longer than 48 hours. The 3,200 ton submarine *Surcouf*, which the French Navy built in 1929, had a crew of 150 and could lie submerged for as long as 120 hours without surfacing.*

The *Surcouff* was able to cruise from France to Madagascar without calling on a single port on the way. As far as all the comforts and amenities were concerned, it probably rivalled Captain Nemo's ship, meanwhile boasting the unquestionable advantage of a waterproof upper-deck hangar for a reconnaissance seaplane. Further note that Jules Verne did not equip his *Nautilus* with a periscope through which the people inside the submarine can view the surface from under the water.

Only on one count will the submarines we make still be a long way inferior to Jules Verne's one: the depth of submergence. However, on this point Jules Verne's imagination stretched beyond the bounds of truth. "Captain Nemo", we read in one place, "went down to 3, 4, 5, 7, 9 and 10 thousand metres below the ocean level." Once the *Nautilus* submerged to the unprecedented depth of 16,000 metres. "I felt," the hero says, "the rivets of the submarine's iron plates shudder and saw its portholes bulge inwards as they yielded to the water's pressure. Had our ship not been as sturdy as a solid cast object, it would have been crushed to pulp instantaneously." He had good reason for such fears because at 16 km down—if such depths really existed—the pressure of the water should reach

$$16{,}000 : 10 = 1{,}600 \ \text{kg/cm}^2$$

---

* Modern nuclear-powered submarines give us a free choice of routes at little explored sea and ocean depths. The inexhaustible stock of power that they carry allow very long cruises without surfacing. Thus, the American nuclear-powered submarine *Nautilus* cruised in Arctic regions without surfacing from the Bering Sea to the Sea of Greenland via the North Pole. Another one of the same class circumnavigated the globe without surfacing once.—*Ed.*

Fig. 49

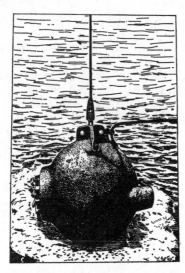

Steel-plated bathysphere in which William Beebe reached the depth of 923 m in 1934.

or 1,600 technical atmospheres. While it wouldn't crush iron, it would unquestionably dent it.

However, oceanographers do not know of such depths. The exaggerated notion of ocean depths, current in Jules Verne's time—the novel was written in 1869—was due to imperfect sounding methods. In those days it was not wire but a hemp rope that was used for a sounding line. The deeper it went, the more friction on water did it experience, until at a certain depth it prevented the line from sinking any further, no matter how much of it was paid out. The rope merely got itself into a tangle and thus produced the false impression of tremendous depths.

Modern submarines can stand a pressure of not more than 25 atmospheres—which means that they can submerge to no deeper than 250 m. Greater depths have been reached in a special diving apparatus called a bathysphere (Fig. 49), which is specially devised for the study of life in the ocean depths. It resembles not Jules Verne's *Nautilus* but rather H. G. Wells' invention in his *The Sea Raiders*, which describes the adventures of a man who went 9 km down to the ocean floor in a thick-walled steel sphere. The apparatus submerged without a cable but with a stock of ballast. On reaching the ocean bed the ballast was jettisoned and the sphere rushed up to the surface. In bathyspheres scientists have reached depths of more than 900 metres. The bathysphere is lowered on the cable from the ship with which the underwater "traveller" maintains contact by telephone.*

How the *Sadko* Was Refloated

Every year, especially in times of war, thousands of ships, big and small, get wrecked. In the last twenty to thirty years the more valuable and accessible of these hulks have been refloated. Soviet engineers and divers of the Special-Purpose Underwater Work Administration have won world fame by successfully raising more than 150 big vessels, largest among which was the icebreaker *Sadko*, wrecked in 1916 in the White Sea due to its skipper's negligence. After 17 years

---

* Later another kind of deep-water diving apparatus called the bathyscaphe was devised in France under the guidance of engineer Willm, and in Italy by the Belgian, Professor Piccard. They differ from bathyspheres in that they can travel at great depths, whereas bathyspheres are strained by the linking cable. At first Piccard went down to more than three kilometres. Then the Frenchmen Guillaume and Willm reached the depth of 4,050 m. In November 1959 a bathyscaphe went down to 5,670 m but even this was not the limit. On January 9, 1960 Piccard went down to 7,300 metres and on January 23 he reached the bottom of the Marianne depression, 11.5 km deep and believed to be the deepest spot in the world.

on the sea bottom this fine ship was raised and refloated.

The technique was based wholly on the rule of Archimedes. At 25 metres down divers drilled twelve tunnels in the sea floor beneath the shipwrecked icebreaker and passed through each sturdy steel cables the ends of which were attached to deliberately submerged pontoons. These pontoons—shown in Fig. 50—

Fig. 50

How the *Sadko* was raised. The icebreaker, the pontoons and the lifting chains are given in cross-section.

were hollow watertight iron cylinders 11 m long and 5.5 m in diameter, each weighing—empty—50 tons, and having a volume of about 250 cu m. Since it is quite plain that an empty pontoon cannot sink—weighing only 50 tons and displacing 250 tons of water it must have a lifting capacity of 250—50=200 tons —it had to be filled with water to submerge.

Aftor the ends of the steel cables were lashed tight to the submerged pontoons (Fig. 50), compressed air was pumped into each of them. At 25 metres down, water exerts a pressure of 25/10+1, i.e., 3.5 atmospheres. Air was pumped in under a pressure of about 4 atm, and, consequently, evacuated the pontoons. The water pushed them up to the surface with tremendous force. The twelve pontoons used had a total lifting capacity of $200 \times 12 = 2,400$ tons, but since this was more than the *Sadko* weighed, not all the water was pumped out—so as to do the job more smoothly. However, the ship was actually surfaced only after several abortive attempts. "We had four failures before we proved successful," T. I. Bobritsky, the engineer in charge of the job, wrote. "Three times, as we waited with bated breath for the ship to appear, we saw instead of the icebreaker the pontoons and the snapped cables and hoses whirling in a chaos of breakers and foam. The icebreaker itself came up and sank twice before it was finally floated."

"Perpetual Motion" Water Machine

Among the countless projects for a "perpetual motion" machine many were based on buoyancy. One was a 20 metre tall tower filled with water, which had a sturdy cable running like an endless belt through

Fig. 51

Project of an imaginary water-driven "perpetual motion" machine.

Fig. 52

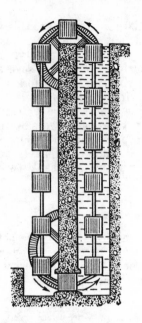

pulleys attached to top and bottom. Fastened to the cable were 14 empty cube-shaped boxes, each a metre high, made of sheets of iron riveted together to prevent water from seeping in. Figs. 51 and 52 give a picture of this tower and its cross-section.

How was it supposed to work? Everyone familiar with the Archimedes principle will gather that in water the boxes try to surface as they are buoyed up by a force equal to the weight of the water they themselves displace—or, in other words, by the weight of 1 cu m of water multiplied by the number of boxes submerged. The figures show us that six boxes are always submerged. Consequently the buoyant force is equal to the weight of six cu m of water, or six tons.

Meanwhile the boxes are pulled down by their own weight, which, however, is offset by the other six cubes freely suspended on the outer side of the cable.

Thus, the cable should be pulled upwards by a force of 6 tons, which should apparently compel the cable to turn endlessly and perform at every turn the work of

$$6,000 \times 20 = 120,000 \text{ kgm.}$$

Hence, if we stud the country with these towers we should be able to make them do an unlimited amount of work—enough at any rate to meet all economic requirements, as they would turn rotors of dynamos and produce any amount of electricity.

However, let us analyse this project. We shall see that the cable will not move at all. Indeed, for the cable to keep on moving, the boxes must dive into the water in the tower from the bottom and re-emerge at the top. But to dive in, each box must *overcome the pressure of a 20-m column of water*. The pressure this column exerts on a square metre of box area is 20 tons —the weight of 20 cu m of water. Meanwhile we have an upward pull of only 6 tons, which is definitely not enough to pull the box into the water.

Among the many water-driven "perpetual motion" machines invented in their hundreds by ill-starred cranks one will find some very simple but witty devices. Fig. 53 shows one of them. Part of the wooden drum shown—which is mounted on an axle—is always submerged. If the Archimedes principle is valid, the submerged segment should always seek to surface; and since the buoying force is greater than the axle friction, the drum ought to spin endlessly. Hold your horses, however! Should you ever try to copy this project, you are bound to fail. The drum simply won't turn. Why? Because the direction in which the forces act has been lost sight of. These forces will always be perpendicular to the drum's surface, i.e., directed radially towards the axle. No doubt you have seen time and again that it is impossible to make a wheel turn by applying an effort radially. Instead, you must apply the effort in a direction perpendicular to the radius—at a tangent to the wheel's circumference.

Fig. 53

Wooden drum

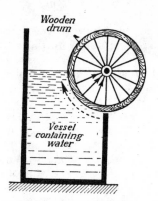

Vessel containing water

Another project for a water-driven "perpetual motion" machine.

Now you see why the attempt to produce "perpetual motion" has again failed.

The Archimedes principle has supplied an endless host of "perpetual motion" cracks with tempting food for thought, inducing them to invent sundry artful devices to exploit the apparent loss of weight so as to obtain a perpetual source of mechanical energy. However, none of these attempts has ever been—nor ever could be—crowned with success.

Who Invented the Word "Gas"?

"Gas" is a word scientists invented along with many other words such as "thermometer", "electricity", "galvanometer", "telephone" and first of all, "atmosphere". Of all invented words, "gas" is unquestionably the shortest. It was Galileo's contemporary the Dutch chemist and physician Helmont (1577-1644) who derived the word from the Greek word "chaos". Having found air to consist of two parts, of which one maintains conbustion and is consumed, while the other doesn't, Helmont wrote: "I have called this vapour *gas*, because it scarcely differs from the *chaos* of the ancients." ("Chaos" originally meant a chasm.) However, for long this newly coined word failed to catch on. It was revived by the famous Lavoisier only in 1789, to gain wide currency when the Montgolfier brothers made their sensational balloon flights.

Lomonosov, the celebrated 18th-century Russian scientist, coined another term for gaseous bodies, he called them "resilient liquids"—a term which incidentally was still in use when I went to school. I might note that Lomonosov introduced into Russian several new standard scientific terms such as "atmosphere", "barometer", "airpump", "crystallisation", "matter", "manometer", "micrometer", "optics", "electricity", "ether", etc. This genius and father of the natural sciences in Russia wrote in this connection:

"I was compelled to look for new words to name certain physical instruments, actions and natural things. These words may seem odd at first but I hope that in time with usage they will grow more familiar."

Lomonosov's prediction, I might add, came true.

A Seemingly Simple Task

A vessel big enough to hold 30 tea-glassfuls of water is filled to the brim. Place a glass under its tap and, watch in hand, note how many seconds it will take to fill up. Suppose it is half a minute. My question now is: how long would it take the vessel to run dry if the tap were turned on?

Doesn't it seem simple? If it takes half a minute for one glass to run out, you may think it should consequently take only a quarter of an hour for the entire vessel to empty.

Try it and see. You'll find that the vessel will empty only in half an hour. How come? After all, it seemed so simple. Yes, but wrong!

Note that the *rate* with which the water runs out is not constant throughout. After the first glass is filled, the second will take longer to fill, because there will be less water in the vessel and, its level having gone down, it will exert a smaller pressure. For the same reason it will take the third glass still longer to fill up—and so on and so forth.

The speed with which any liquid pours out of a hole in the wall of an open-mouthed vessel with an open top is in direct proportion to the height of the column of liquid above the hole. Torricelli, Galileo's brilliant pupil, was the first to note this dependence which he expressed in the simple formula: $v = \sqrt{2gh}$, in which v is the speed with which the liquid pours out, g is the acceleration of gravity, and h is the height of the column of liquid above the orifice. From this formula it follows that the speed with which the liquid spills out does not at all depend on its *density*. When the column height is the same, light alcohol and heavy mercury will pour out with the same speed (Fig. 54). Furthermore on the moon, whose gravity is but a sixth of the earth's, it will take roughly two and a half times more to fill a glass than here on earth.

However, back to our problem. If after the vessel feeds twenty glasses, the water level—counting from the tap—drops to a *quarter* of the full vessel, it will take the 21st glass *twice* as long to fill up as the first glass. If later on the level of the water is down to a ninth, it will take already *thrice* as much time to fill up the last few glasses. Using the calculus to solve the problem, we find that the time required to let the ves-

Fig. 54

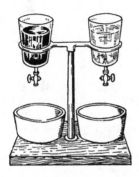

Which of the two—mercury or alcohol—will pour out sooner? The levels in both vessels are identical.

sel run dry is *twice* as much as the time needed for the same amount of liquid to pour out were the level of water above the top to remain constant.

The Tank Problem

From what I have just said it is one step to the notorious tank problem that one finds in every single collection of arithmetical and algebraic problems. No doubt, you all remember those classic scholastically dry problems of this order:

"The tank has two pipes—one leading in and the other out The first needs five hours to fill the tank to the brim, and the second ten hours to drain it dry. How long will it take the tank to fill up when the stopcocks are out of both pipes?"

This problem has a venerable history, dating right

back some twenty centuries to Heron of Alexandria. Here is one of his problems,

> *Four fountains are there and a reservoir vast.*
> *In but one day the first doth fill it to the brim.*
> *The second two days and nights must play to do the*
> *same.*
> *The third takes thrice the time as did the first.*
> *The fourth comes last with four days and nights.*
> *Now tell me when the reservoir will fill,*
> *When all four play at once.*

It is two thousand years now that the tank problem has been posed and—such is the force of habit!—has been solved *wrongly* in all this time. After our tank problem, you should realise why. Indeed, what is the

Fig. 55

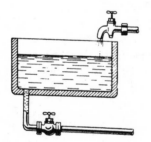

The tank problem.

solution given? The following for the tank one mentioned above. In one hour it is claimed the first pipe fills a fifth of the tank, while the second drains a tenth of it. Consequently, with both open the tub should fill up every hour by 1/5—1/10=1/10, which means that it should take ten hours for the tank to fill to the brim. This is the wrong way of going about it, though. Whereas the water may be considered to be flowing in under a constant and consequently uniform pressure, it *flows out* under the pressure exerted by a changing level; hence the flow is *not uniform*. From the fact that it takes the second pipe ten hours to drain the tank it does not at all follow that a tenth of the water in the tank flows out with every hour. We shall never solve this problem correctly with the help of elementary mathematics, and so all problems dealing with tanks and the *outflow* of water really have no place in collections of arithmetical problems.

Amazing Vessel

Could we have a vessel from which water would keep flowing out evenly without slowing down to a trickle, even though its level inside would drop? I

Fig. 56

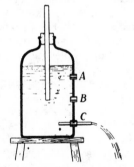

Cross-section of "Mariotte's bottle". The water flows out in a steady jet.

suppose that now you will think this impossible to accomplish. But it can be done. The bottle depicted in Fig 56 is one such amazing vessel. It is an ordinary bottle with a narrow mouth, having a glass tube inserted through the cork. When you open tap *C*, below the end of the tube, the liquid flows out evenly until the level of the water inside the vessel drops below the end of the tube. By pushing the tube down almost to tap level you can cause all the liquid above the tap to flow out uniformly—even though in a mere trickle.

How does this come about? Try to trace in the mind's eye what happens in the vessel when you open tap C to let the water out. It is first of all the level of the water in the top tube that falls to the bottom of it; only afterwards does the level of the water in the vessel itself descend and outside air enters through the now empty tube. As the water flows out, its level in the vessel drops. Meanwhile the air from outside enters through the glass tube to join the rarefied air beneath the water, bubbles up and gathers above the water in the upper part of the bottle. Now at the entire level of tap B, the pressure is equal to atmospheric pressure. Consequently the water flows out through tap C merely owing to the pressure exerted by the BC layer of water, since atmospheric pressure both inside and outside the bottle is identical. As the height of the BC layer stays constant, no wonder the water flows out with a uniform speed all the time.

Now try to answer this question. How quickly will the water flow out if plug B—which is level with the tube end—is pulled out? Surprisingly enough, we find that *it won't flow out at all*—provided, of course, the hole is small enough to be disregarded, because otherwise the water will flow out due to the pressure of the thin upper layer which is just as high as the hole is wide. Actually both inside and outside the pressure is the same as atmospheric pressure and so there is nothing to cause the water to flow out. But if we pull out plug A, which is *above* the end of the tube, we shall see the outer air entering the bottle instead of water flowing out. Why? For the very simple reason that inside this part of the vessel the pressure of the air is *less* than the outside atmospheric pressure.

132

This vessel, which has such unusual properties, was invented by the famous physicist Mariotte and has hence come to be known as "Mariotte's bottle."

Load of Air

In the mid-seventeenth century, the citizens of Regensburg and the German princes with the Emperor at their head, witnessed the following wonder. Sixteen horses in teams of eight apiece tried might and main to pull apart two copper hemispheres joined together by "nothingness"—air itself—and failed. So did Burgomaster Otto von Guericke, the "German Galileo" as he is sometimes called, show for everyone to see that air was not at all "nothing", that it possessed weight and that it exerted a considerable pressure on everything on earth.

This experiment was staged in great ceremony on May 8th, 1654. The learned burgomaster was able to interest everyone in his studies, despite the political confusion and devastating wars at the time.

Though you will find a description of it in practically every textbook on physics, I'm sure you won't be averse to hearing the story from Guericke himself. A fat volume describing a long series of his experiments was put in Latin in Amsterdam in 1672. Like all books of the period it had a very rambling title. Here it is:

OTTO von GUERICKE

The so-called new Magdeburg experiments with

AIRLESS SPACE

originally described by KASPAR SCHOTT,

Professor of Mathematics of Würzburg University.
Published by the author himself,
in a more comprehensive form and supplemented with
various new experiments.

It is Chapter 23 that deals with the experiment mentioned. Here it is:

"An experiment demonstrating that the pressure of air can join two hemispheres so firmly together, that even sixteen horses will not pull them apart.

"I ordered two copper hemispheres of three quarters of a Magdeburg ell [equal to 550 mm] across. Actually they were only 67/100 ell across, since the craftsmen, as is their habit, could not make exactly what was required. The two hemispheres were quite identical. One had a stopcock to evacuate the inside air and prevent the outside air from entering. The two hemispheres also had four rings attached to lash on the ropes fastened to the harnesses of the horses. I also ordered that a leather ring be made which I soaked in a mixture of paraffine and turpentine. This I placed between the two hemispheres to prevent any air from entering. The nozzle of an air pump was then inserted in the stopcock, and the air inside the sphere was evacuated. The force with which the two hemispheres adhered to each other through the leather ring was then revealed. The pressure of the outer air glued them so tightly together that sixteen horses failed to pull them apart, or did so with great difficulty. When the hemispheres yielded to the horses' pull and came apart, there was a loud report as if a shot had been fired. However, one turn of the stopcock to let the air enter freely was enough to easily pull the hemispheres apart with one's hands."

A simple reckoning will show why so great a force (eight horses from each side) had to be exerted to pull the two halves of the empty sphere apart. Air exerts a pressure of about 1 kg/cm². The area of a circle with a diameter of 0.67 ell (37 cm) is 1,060 cm². (We take the *area of a circle* and *not the surface of the hemisphe-*
re, because atmospheric pressure is equal to the indi-

cated value only when it acts on a surface at right angles, being less for an inclined surface. In the case in question we take the right-angled *projection* of the sphere's surface, or, in other words, the area of the great circle.) Consequently, atmospheric pressure on each hemisphere should exceed 1,000 kg or one ton. This means the two eight-horse teams must have pulled with a one-ton force to counteract the pressure of the outer air.

Now one ton does not seem so very much for so many horses. However, don't forget that when horses pull a one-ton load, they have to overcome a force not of one ton, but much less. This force, the friction between the wheels and the axles and between them and the road, amounts on a highway, for example, to only 5 per cent of the load hauled—which in the case of a one-ton load would mean only 50 kg (let alone the fact that half of the pulling force is lost, as practice has shown, when eight horses pull together). Consequently, for eight horses a one-ton pulling force is tantamount to a 20-ton cart. That is the load of air which the Magdeburg burgomaster's horses had to pull. They had to get moving what seemed to weigh as much as a small locomotive—moreover one not on rails.

A dray horse can exert a pulling force of about 80 kg (when moving at the speed of 4 km/h). A horse exerts a pulling force averaging 15 per cent of its weight. A race-horse weighs some 400 kg and a dray horse some 750 kg. For a very short time—the initial effort—the pulling force may be several times greater. Consequently, to pull the Magdeburg hemispheres apart, we should need 1,000:80=13 horses on either side.

You may be surprised to learn that some of the joints of our skeleton adhere precisely because of this same reason. Our pelvis is a beautiful illustration to the Magdeburg hemispheres. Even if we stripped it of the muscle and gristle binding it together; it would still stay in place. It is atmospheric pressure

Fig. 57

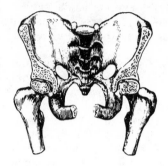

Atmospheric pressure keeps the human pelvis together— just as it kept the Magdeburg hemispheres together.

which does that—there is no air in the spaces between the joints.

Modifications of Heron's Fountain

You most likely know what the usual kind of fountain—ascribed to the ancient mathematician Heron of Alexandria—looks like. In any case I shall remind you of its main points before describing the latest modifications of this curious device. Heron's fountain (Fig. 58) consists of three vessels—the upper one, *a*, is an open plate-shaped affair, while *b* and *c* are two air-tight spherically-shaped retorts. All three are connected by three tubes as shown in the figure. The fountain begins to play when vessel *a* has a little water in it, vessel *b* is full of water, and vessel *c* is full of air. The water flows along the tube from vessel *a* to vessel *c* forcing the air into vessel *b*. The pressure of this incoming air causes the water from vessel *b* to rush up the tube and play above vessel *a*. As soon as all the water runs out of vessel *b*, the foun-

Fig. 58

Cross-section of Heron's fountain.

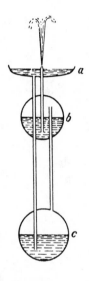

tain ceases to play. That was how the fountain worked in Heron's times.

More recently, one Italian physics master, whose school lab's meagre furnishings prodded him to display ingenuity, simplified the fountain and introduced modifications which every one of you could easily reproduce with the simplest of means (Fig. 59). In place of spherical retorts and glass or metal pipes he used flasks and rubber tubes. The upper vessel does not have to have a hole in the bottom. The tubes can be made to dangle over the rim as shown at the top of Fig. 59.

Fig. 59

Cross-section of a modern modification of Heron's fountain. Top: variant for the upper vessel.

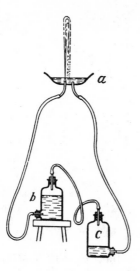

This is far more convenient to use, because after all the water from flask *b* has run via vessel *a* into flask *c*, you may simply interchange *b* and *c* and your fountain will begin to play again. However, when you do that, don't forget to reset the nozzle. Another convenience is that we may rearrange the vessels at will to see how their different levels affect the height of the spray.

Should you want to increase the height of the fountain by a number of times, all you have to do is to fill the two flasks with mercury and water, in place of water and air (Fig. 60). You will clearly see that

Fig. 60

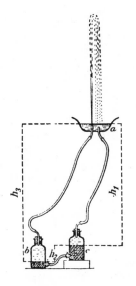

Cross-section of a mercury-pressure fountain. The jet shoots up to a height ten times greater than the difference between the mercury levels.

as it flows from c to b the mercury will oust the water in a fountain. Knowing that mercury is 13.5 times heavier than water, we may reckon the height of the fountain. Designate the different levels as respectively h_1, h_2 and h_3. Let us now see what forces the mercury to flow from c (Fig. 60) to b. The mercury in the connecting tube is subjected to pressure from two sides. One is the pressure exerted by h_2, the difference between the levels of the columns of mercury (which is tantamount to the pressure of a column of water 13.5 times taller, that is 13.5 h_2) plus the pressure exerted by the column of water h_1. Such is the pressure exerted from the right. On the left a pressure is exerted by the column of water h_3. As a result there acts upon the mercury a force of 13.5 $h_2 + h_1 - h_3$. But since $h_3 - h_1 = h_2$, we replace $h_1 - h_3$ by $-h_2$ to get 13.5 $h_2 - h_2$, i.e., 12.5 h_2. Hence the mercury is forced into flask b by the pressure of a column of water equal in height to 12.5 h_2. Theoretically the fountain should thus play to a height equal to the difference between the mercury levels of the two flasks multiplied by 12.5. However, because of friction the actual height is a little less.

However, this device conveniently affords a rather high fountain. To get a 10-metre height it is quite enough to place one of the flasks roughly a metre higher than the other. Curiously enough the raising of vessel a further up above the flasks with the mercury does not at all affect, as our reckoning shows, the height of the fountain.

"Don't Get Wet"

Back in the 17th and 18th centuries, noblemen amused themselves with the following instructive toy. This was either a mug or a small pitcher with a fanciful pattern of big holes at the top (Fig. 61). Aristocrats would offer such a mug full of wine to a guest from the lower classes, at whose expense a laugh could be had with impunity. And, indeed, how is one to drink from it? One can't tip it because then the wine would flow out through the many holes in the top and not a drop would reach one's lips.

However, all who knew the secret—shown in Fig. 61 on the right—stopped up hole B with a finger, and sucked in the liquid through the spout without

Fig. 61

"Don't get wet" 18th-century drinking vessel and its secret.

tilting the vessel. In this way the wine rises up through hole E, up the channel inside the handle and thence along its continuation C inside the upper rim of the mug till it reaches the spout. These potter's jokes are made in the Soviet Union even today. I myself have seen some of them, with the words "Drink, but don't get wet" inscribed on them.

How Much Will the Water in an Overturned Glass Weigh?

Nothing, you will say, as the water will surely pour out. But suppose it doesn't? We can make water stay in an overturned glass without pouring out. Fig. 62 shows you how this is done. The overturned wineglass, attached to the bottom of one of the pans, is full of water which does not pour out, as the rim of the glass is immersed in a jar of water. The other pan holds an empty wineglass of identical shape. Which is the heavier of the two? That to which the overturned water-filled wineglass is attached, since, while subjected to full atmospheric pressure from the

Fig. 62

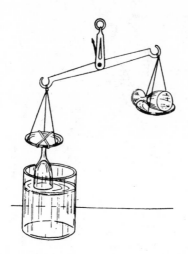

top, from the bottom it is subjected to atmospheric pressure which has been reduced by an amount equivalent to the weight of the water in the wineglass. To make the scales balance we must fill up the glass on the other pan with water. Consequently the water in the overturned glass weighs as much as it would in a glass standing upright.

Why Do Ships Attract One Another?

In the autumn of 1912, the ocean liner *Olympic*, one of the world's biggest ships at the time, was steaming ahead out on the high seas, when another much smaller ship, the cruiser *Hawk*, rapidly approached it on a parallel course a hundred metres away. As soon as the two ships took up positions as depicted in Fig. 63 a surprising thing happened. The

142

Hawk sharply veered off its course, as if obeying some invisible force, turned towards the big liner and heedless of the helm rammed into it. The impact was so great that it made a big gash in the *Olympic*'s hull. A tribunal examined this queer case and found the *Olympic*'s skipper guilty, as, according to its ruling, he had failed to issue orders to yield the right of

Fig. 63

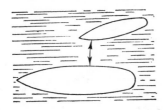

Diagram shows the positions of the *Olympic* and *Hawk* before collision.

passage to the Hawk. Consequently, the tribunal, as you can gather, saw nothing extraordinary about it all, attributing the accident to the skipper's negligence. Actually this was the result of a totally unforeseen circumstance, a case of the mutual attraction of ships at sea.

This must have taken place before, too, with two ships on a parallel course. But when ships were small, their mutual attraction was not so much in evidence. However, now that floating cities are plying the oceans, it is far more noticeable and warship commanders duly reckon with it when manoeuvring. It was most likely to blame for the many smaller boats that collided with bigger ships, when cruising in their vicinity.

What causes this attraction? Newton's law of universal gravitation has, of course, nothing at all to do with it. We already know from Chapter 4 that this attraction is negligible. There is a totally di-

143

fferent cause: it derives from the laws that govern the flow of fluids in pipes and channels. One can prove that in the case of a fluid flowing through a channel with narrower and wider parts, it will flow faster through the narrower part and exert a smaller pressure on the channel's walls than in the wider parts, where its flow is calmer and its pressure on the walls greater. This is called the *Bernoulli principle*.

The same holds for *gases*—called in this case, however, the *Clément-Desormes effect*, after the physi-

Fig. 64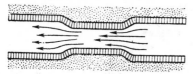

In the narrower cross-section the water flows faster and exerts a smaller pressure on the walls of the channel.

cists who discovered it and often known as the "aerostatical paradox". It is said that this phenomenon was discovered accidentally in the following circumstances. At one French mine, a collier was told to cover up with a board the outlet of a shaft through which the compressed air was fed into the mine. The miner had a hard job battling with the jet of air coming out of the shaft. All of a sudden the board banged the shaft tight so violently that had it not been large enough, it would have been pulled into the ventilation shaft together with the scared workman. Incidentally this characteristic feature of the flow of gases explains how an atomiser works.

When we blow into pipe *a* (Fig. 65), which tapers off into a nozzle, the pressure of the air in it decreases. As a result above nozzle *b* air pressure is less: that is why atmospheric pressure forces the liquid in the glass to rush up the pipe. As it comes out of the nozzle, the liquid is atomised in the jet of air.

Now we shall understand why ships attract each other. In the case of two ships on a parallel course, we have a channel of water between them, the difference being that whereas in an ordinary channel the walls remain stationary and the water moves, here the water remains "stationary" and the "walls" move.

Fig. 65

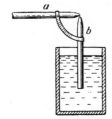

This difference, however, does not at all change the action of the forces. In the narrower parts of the moving channel the water exerts a smaller pressure on the "walls" than elsewhere. In other words the sides of the ships facing each other are subjected to a smaller pressure than the two outer sides. What happens then? The pressure of the water on the outer sides causes the ships to move towards each other, with the smaller boat, naturally, moving much faster. The bigger boat hardly seems to move. That is why attraction is so strong when a big boat rapidly bypasses a small one.

So to sum up: the mutual attraction of ships is due to the suction of flowing water. This also explains the danger that rapids and the suction of whirlpools present to bathers. It has been reckoned that a stream of water, moving with the moderate speed

145

of one metre a second, attracts a person with the force of 30 kg. It is not so easy to resist so strong a force especially in water, when your own weight prevents you from keeping your balance. Finally, the suction of a swift train is also due to the same Bernoulli principle. A train dashing along with the speed of 50 km/h will attract a bystander with a force of about 8 kg.

Fig. 66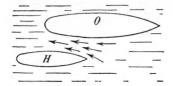

Flows between two ships in motion.

The layman knows very little about the phenomena associated with the Bernoulli principle—though they are none so seldom. Therefore I think it would be of benefit to dwell on this in greater detail. The following is an extract from an article on the subject written for a popular-science magazine.

The Bernoulli Principle and Its Consequences

This principle, first stated by Bernoulli in 1726, is: in an air or water flow pressure is great when velocity is small and vice versa. There are, of course, some limitations but we shall not deal with them here.

Fig. 67 illustrates the principle.

Air is blown in via tube AB. In the "bottleneck" a the velocity of the air is great, in the broader part b it is small. Where the velocity is great, pressure is small and vice versa. Owing to the lower pressure at point *a*, the liquid in tube *C* rises, meanwhile, the increased pressure at point *b* forces down the liquid in tube *D*.

In Fig. 68 tube *T* is mounted on disc *DD*; air is blown in via *T* and further past the nonfastened disc *dd* (to simplify the experiment take a used spool and

Fig. 67

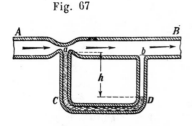

Illustration of the Bernoulli principle. In the "bottleneck" (*a*) of tube *AB* pressure is less than in the broader part (*b*).

Fig. 68

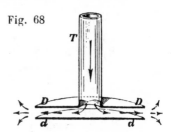

The discs experiment.

a paper disc, kept in position by means of a pin inserted into the spool's hole). The air between the two discs possesses great velocity which rapidly diminishes the nearer it is to the rims, since the cross-section of the air flow quickly grows and the inertia of the air flowing out from in between the discs has to be overcome. However, since the pressure exerted by the air around the disc is great because its velocity is small, while the pressure exerted by the air between the discs is small, because its velocity, on the contrary, is great, the pressure of the outer air, being the greater, forces the discs together with an intensity that is the more powerful, the more intensively air is blown in through *T*.

Fig. 69 is practically identical with Fig. 68, only it deals with water. The fast moving water on top

Fig. 69

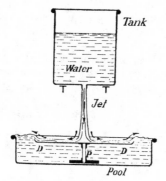

Tank

Water

Jet

Disc *DD* rises along rod *P* when a jet of water comes down on it.

D — D

P

Pool

of disc *DD* maintains a lower level, rising to the higher level of the placid water in the tank only when overlapping the disc. The placid water, hence, exerts a greater pressure than the moving water on top the disc, and as a result the disc rises. (The disc is kept in position by rod *P*.)

Fig. 70 depicts a small pith ball floating in a jet of air. The air strikes at the ball and prevents it from

Fig. 70

Air jet supports the little ball in mid air.

falling: Meanwhile, should the ball pop aside, the outer air—whose pressure is greater since its velocity is smaller—returns it to the jet.

Fig. 71 depicts two ships on a parallel course in placid water, or standing still in flowing water—which amounts to the same thing. In between the vessels the velocity of the narrower flow is greater

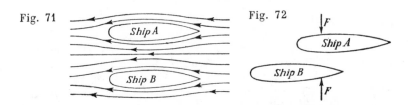

Fig. 71

Two ships moving in parallel directions attract each other.

Fig. 72

When the ships move forward, ship *B* veers nosewards towards ship *A*.

than that of the outer water and, consequently, the pressure exerted by the water in this place is less, with the result that the greater pressure of the outer water compels the two ships to draw together—a phenomenon well known to seamen.

Fig. 72 gives a more serious case when one ship is a bit in advance of the other. The two forces *F* and *F* causing the ships to move together, try to turn them, with *B* veering around towards *A* with an appreciable force. A collision is inevitable as the movement is too rapid for the helmsman to manage to change the course.

One can demonstrate the phenomenon illustrated in Fig. 71 also by blowing air between two light rubber balls suspended as in Fig. 73. When this is done, they swing into contact.

149

Fig. 73

When a jet of air is blown between these two light balls they swing into contact.

Why Fish Have Bladders

What does the bladder do for a fish? It is usually said—and, as one would think, with good reason—that the fish inflates its bladder to ascend to the surface: this supposedly increases the body volume and thus the weight of the water displaced presumably becomes greater than the fish's own weight. Buoyancy causes the fish to rise. When the fish wants to stop ascending, or to descend, it supposedly compresses its bladder, thus reducing its body volume and, consequently, the volume of the water displaced, and sinks again in conformity with the Archimedes principle.

This volgarised notion of the purport of the fish's bladder dates back to the 17th century and was first postulated in 1675 by Prof. Borelli of the Florentine Academy. For more than 200 years it was accepted unreservedly, and thus set out in every school book. Only thanks to recent investigations was it completely disproved.

The bladder undoubtedly helps the fish to swim, as the fish from which it had been removed were able to keep afloat only by dint of hard work with their fins; as soon as they stopped they plummeted to the bottom like stones.

So what is the bladder's real purpose then? It

plays a very limited role. All it does is to help the fish stay at a certain depth, where the weight of the water displaced is equal to the weight of the fish itself. When the fish, by moving its fins, *descends*, its body together with the bladder is compressed by the great pressure of the surrounding water. This reduces the weight of the water displaced to less than that of the fish itself and thus the fish sinks. The lower it goes the greater the pressure becomes, rising by one atmosphere with every ten metres. This compresses the fish's body still more, making the fish sink more rapidly.

The same process, only in reverse, takes place when the fish lifts itself up by fin-work from the layer of water where it was in a state of equilibrium. Now surrounding pressure diminishes; but the fish's bladder—wherein the pressure offset that of the surrounding water—causes its body volume to increase and thus be buoyed up. The higher up the fish goes the more bloated its body becomes, and the more rapidly does it ascend. The fish cannot halt its ascent by "compressing its bladder", as the walls of this bladder have no muscles.

The following experiment has confirmed the fact of *passive* expansion of a fish's body volume (Fig. 74). A chloroformed bleak was placed in a sealed water-filled jar in which an intensified pressure, approaching that at a certain depth in a natural body of water, was maintained. On the surface the fish lay passively with its belly up. When pushed a little down, it again floated up to the top. When pushed down nearer to the bottom, it sank right to the bottom. In between these two levels, it remained in a state of equilibrium, neither sinking further nor surfacing. Recall what I have just told you about the passive expansion and compression of the bladder and you will realise why

Fig. 74

So, in defiance of the common notion, a fish can't compress or expand its bladder at will. Its volume varies but does so passively, due to increased or reduced outer pressure—in conformity with the Boyle-Mariotte law. This varying volume only harms the fish as it makes it either rush faster and faster up or dive rapidly down. In short, the bladder helps the fish to preserve its equilibrium when it isn't moving, but the equilibrium is *unstable*. Such is the true role of the fish's bladder—as far as swimming is concerned. We don't know whether it has any other functions—so far the bladder is still a mystery. It is only its hydrostatic role that has been satisfactorily explained so far.

Fishermen's observations confirm what I have said. It sometimes happens in deep-water fishing that the fish manages to slip off the hook or out of the net. However, contrary to expectation, it doesn't dive back to where it was caught but rushes up, very often with the bladder jutting out of its mouth.

Waves and Whirls

The elementary laws of physics fail to explain many common physical processes. Even such a frequently observed phenomenon as sea waves on a windy day finds no full explanation within the limits of school physics. What causes waves in calm water when a boat cuts through it? Why do flags flutter on a breezy day? Why is seaside sand wave-shaped? Why does the smoke from a chimney stack curl?

Fig. 75   Fig. 76

The calm, laminar flow of a fluid in a pipe.   The disquiet, turbulent flow of a fluid in a pipe

To explain all these and other similar phenomena we must know the peculiar features of *whirling* flow of fluids and gases. Since school textbooks hardly mention this at all, here are a few salient points.

Imagine a fluid flowing through a pipe. When all the particles in the fluid move through the pipe along parallel lines, we have the simplest type of fluid flow—the calm or *laminary* flow as physicists call it. However, this is not so frequent. On the contrary, far oftener does a fluid flow through pipes in a disquiet state, with vortices spreading from the pipe's walls to its axis. This is the vortical or *turbulent* flow which we observe in water-mains—provided we discount thin pipes where we have a laminar flow. This turbulent flow is observed every time the velocity of the flow of the fluid in question in a pipe of a given diameter reaches a certain magnitude called the *critical* velocity. (The critical velocity for a definite fluid is directly proportional to its viscosity and inversely proportional to its density and the diameter of the pipe through which it is flowing.)

We can render observable the vortices which a liquid makes while flowing through a pipe by adding to a transparent fluid in a glass pipe some amount of light powder, such as lycopodium powder, for instance. We shall then be able to clearly discern the vortices spreading from the pipe's walls to its axis.

This characteristic feature of the turbulent flow is used to advantage in refrigerators and freezers. A fluid moving in a turbulent flow through a pipe with cooled walls brings all its particles into contact with these cold walls much sooner than if it were flowing otherwise. One must realise that fluids as such are poor conductors of heat and, in the absence of intermixing, cool or warm up very slowly. The blood and the tissues around it are able to exchange heat and substances so vigorously precisely because the flow of blood through the blood vessels is turbulent and not laminar.

All that has been said of fluids in pipes goes for open canals and streams too, as here the water moves turbulently. When we take exact measurements of the velocity of a river's flow, our instruments register pulsations, especially near the bottom, which point to a constantly changing direction of the flow, or, in other words, to turbulence. The water particles in a river move not only along its course, as is usually thought, but also away from its banks to the middle. Therefore to claim that deep down in a river the water maintains a constant round-the-year temperature of 4°C above zero is wrong. Due to intermixing, the temperature of the flowing water near the bottom of a river—but not a lake—is the same as at the surface.

The vortical flow at the river bottom carries away the light sand and produces "waves" in the sand. You will see the same on a beach paved by the tide (Fig. 77). Were the near-bottom flow steady the sandy bed would be even.

Fig. 77

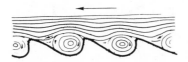

Turbulence causes sand waves on a sea beach.

So to sum up: we shall find a turbulent flow on the surface of all bodies washed by water. That it exists is illustrated, for instance, by the snaking motions of a rope, of which one end is tied fast and the other dangling freely in the direction of the flow. How is this motion produced? A vortex arises near a section of the rope and pulls at it. The very next instant, another vortex appears to pull at it in the opposite direction, thus combining with the first to produce the snaking motion (Fig. 78).

Let us now turn from fluids to gases; from water to air. No doubt you have seen whirlwinds snatch up dust and straw from the ground. That is a manifes-

Fig. 78

Turbulence causes a rope to snake in flowing water.

tation of the turbulent flow of the air near the ground. When the air flows along the surface of water, "humps" appear, causing waves or ripples wherever vortices arise due to a drop in atmospheric pressure. The same thing is responsible for the sand "waves" in a desert or on the dune slopes (Fig. 80).

Now you will understand why a flag flutters in a breeze. This is identical with the rope-in-the-water case. The weather-cock never points in one constant direction on a windy day, but swivels obedient to

Fig. 79

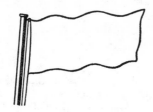

Flag fluttering in a breeze.

air turbulence. The clouds of smoke emerging from a factory chimney stack are of a similar turbulent origin. The gases from the furnace swirl up the chimney and continue to whirl by inertia for some time after they escape from the chimney (Fig. 81).

Air turbulence is extremely important for aircraft. Airplane wings are shaped so as to fill up the rarefied air space beneath and enhance the vortices above. This produces a combination of buoyancy (from below) and suction (from above), Fig. 82. This is similar to what takes place when birds soar with outspread wings.

What does the wind do, when it blows against a roof? Its vortical motions cause the air above the roof to rarefy, and in trying to offset pressure, the air, beneath the roof, presses on it. As a result we often see the wind carrying away a light, loosely fastened roof. This also explains why on a very windy day, large shop windows bulge out due to inside pressure. (They are not shattered by outside pressure.)

Fig. 80

Desert "ripple".

Fig. 81

Smoke belching from a
chimney stack.

There is, however, a simpler explanation for these
phenomena. They are due to the reduction of pressure in air flows. (See above "The Bernoulli Prin-

ciple".) When two air flows of different temperatures and humidities take parallel courses, vortices arise in each. Incidentally, this is the main reason for the

Fig. 82

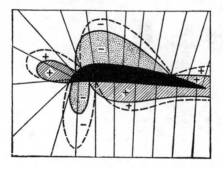

Diagram showing the forces acting on the wing of an airplane. Experiment-proved pattern of the distribution of air pressure (+) and rarefications (—) on the wing. The combined application of the forces of buoyancy and suction gives the wing a lift. (The solid lines show the distribution of pressure; the broken lines show the same when flight speed is greatly increased.)

different shapes clouds assume. This all shows you how wide is the range of phenomena associated with vortical flows.

Journey to the Centre of the Earth

Nobody has ever descended to more than 3.3 km — though the earth's radius is 6,400 km and there is still a long way to go to its centre. Only that ingenious science-fiction novelist, Jules Verne, was able to send his eccentric professor with his nephew way down towards the centre of the earth. The amazing adventures of the subterranean travellers are described in *Journey to the Centre of the Earth*. One of

158

the many unexpected difficulties that the professor and his nephew had to cope with was, incidentally, increasing air density. You know that the higher up one goes, the more rarefied the air becomes, its density diminishing in geometric progression, with height increasing in arithmetic progression. On the contrary, the lower one descends beneath ocean level, the denser the air becomes under the pressure of the abovelying layers. The professor and his nephew couldn't help noticing this, naturally, and this is what they had to say to each other after they reached the depth of 12 leagues or 48 km.

"'Now,' he said, 'consult the manometer. What does it indicate?'

"'Considerable pressure.'

"'Well, then, you see in descending gradually we get accustomed to the density of the atmosphere, and are not the least affected by it.'

"'Not in the least, except a little pain in the ears.'

"'That is nothing, and you can get rid of it at once by breathing very quickly for a minute.'

"'Quite so,' said I, determined not to contradict him again. 'There is a positive pleasure even in feeling one's self getting into a denser atmosphere. Have you noticed the wonderful clearness of sound here?'

"'Yes, indeed. A deaf man would soon get his hearing again.'

"'But this density will of course increase?'

"'Yes, according to a somewhat indefinite law. It is true that the intensity of the weight will diminish in proportion as we descend. You know that it is on the surface that its action is most felt, and at the centre of the globe objects have no longer any weight?'

"'I know that; but tell me, in the end will not the air acquire the density of water?'

"'Undoubtedly, under the pressure of 710 atmospheres.'

"'And lower still?'

"'Lower still of course the density will increase still more.'

"'How shall we descend, then?'

"'Well, we must put stones in our pockets.'

"'I declare, uncle, you have an answer for everything.'

"I did not dare to go any further into the field of hypothesis, for I should have been sure to have stumbled against some impossibility, which would have made the professor start out again.

"But it was quite evident that the air, under a pressure of possibly a thousand atmospheres, would pass at last into a solid state; and in that case, even supposing that our bodies might have held out, we should be forced to stop in spite of all the reasonings in the world."

Imagination and Mathematics

Let us verify Jules Verne's statement. We shall see that the novelist was wrong and we won't have to descend into the bowels of the earth either to discover his mistake. All we must do is to equip ourselves with a pencil and a piece of paper.

Firstly, let us try to find how far down we must go for atmospheric pressure to increase by a thousandth. Normal atmospheric pressure is always equal to the weight of a 760-mm column of mercury. If we were living in mercury we would have to descend by only $\frac{760}{1,000} = 0.76$ mm for pressure to increase by a thousandth. Actually, we shall have to descend much further, as many times deeper as air is lighter than mercury, or

10,500 times more, to be exact. So for pressure to increase by a thousandth we must descend not 0.76 mm (as in mercury) but $0.76 \times 10,500$, almost 8 metres. With every 8 metres down, pressure will increase by one-thousandth (since each subsequent 8-metre air layer will be denser than the previous one and the absolute physical increase in pressure will also be greater than in the foregoing layer—which is what it should really be as this is the thousandth of a larger amount). Wherever we would be—at the top of the world (22 km), on Mount Everest (9 km) or near sea level—we must make an 8-metre descent for atmospheric pressure to increase by a thousandth. This will give us the following table showing how atmospheric pressure increases with depth:

at ground level, pressure $= 760$ mm $=$ normal
at 8 m down, pressure $= 1.001$ of normal
at 2×8 m down, pressure $= (1.001)^2$ of normal
at 3×8 m down, pressure $= (1.001)^3$ of normal
at 4×8 m down, pressure $= (1.001)^4$ of normal

So, at $n \times 8$ m down, atmospheric pressure will be greater than the normal by $(1.001)^n$ times; and while pressure is not very great, air density will increase at the same rate (the law of Mariotte).

Now Jules Verne tells us that the professor and his nephew descended only to 48 km. Consequently, we may discount the decrease in gravity and the related reduction in the weight of the air. So how great was the pressure that Jules Verne's subterranean travellers experienced at the depth of 48 km or 48,000 m? In our formula n is equal to $48,000 : 8 = 6,000$. We must therefore calculate $1.001^{6,000}$. Since the multiplication of 1.001 by itself 6,000 times is a rather tedious and time-consuming task we shall address ourselves to logar-

161

ithms—of which the great French astronomer Laplace justly said that by cutting the amount of labour consumed they double the lifetime of the reckoner. (Any of you who dislike logarithms may change your mind after you read the following extract from Laplace's *Exposition du système du monde*: "The invention of logarithms, reducing several months of reckoning to a few days of work, doubles as it were the lifetime of an astronomer, saving him from the fallacies and fatigue always attendant upon long calculations. The human mind may pride itself upon this achievement, all the more so, since it derives wholly from the human mind. In technology, man draws upon the materials and forces of surrounding nature to multiply his power. Logarithms, however, are the product solely of his own mind.") Using logarithms, we find

$$\log x = 6{,}000 \times \log 1.001 = 6{,}000 \times 0.00043 = 2.6.$$

From the logarithm of 2.6 we learn that x is equal to 400.

So at the depth of 48 km atmospheric pressure is 400 times greater than normal. Under this pressure the density of the air would increase, as experiments have shown, by 315 times. We can hardly believe, therefore, that Jules Verne's subterranean travellers felt nothing more than "a pain in the ears". However, Jules Verne goes on to say that they went down still further—to the depth of 120 km and even 325 km. At such depths atmospheric pressure must have reached monstrous proportions. Meanwhile, we know that human beings can bear without harm an atmospheric pressure of not more than 3 to 4 atmospheres.

Using the same formula to estimate the depth at which air would become as dense as water, that is 770 times denser, we would get 53 km. This would be wrong, however, for in the realm of big pressures the density of gases is no longer in direct proportion to pressure, Mariotte's law is valid only for pressures of not more than several hundred atmospheres. Here is a table of experimentally demonstrated air densities:

	Pressure	Density
200	atmospheres	190
400	"	315
600	"	387
1,500	"	513
1,800	"	540
2,100	"	564

As we see, the increase in density quite noticeably lags behind the increase in pressure. In vain did Jules Verne's professor believe that he would reach a depth where air would be denser than water. He would have never reached that depth, since air becomes as dense as water only under a pressure of 3,000 atmospheres. Beyond that pressure, it scarcely compresses at all. Moreover, we shall never be able to solidify air by pressure alone, unless we "freeze" it to a temperature of 146°C below zero.

To be just, one must note that Jules Verne published this novel long before the facts I have given you were made known. But though this exonerates the author, it doesn't make his story true.

Let us draw once again on the above-mentioned formula to estimate how far down a human being could go without doing himself any harm. The greatest atmospheric pressure we can stand is 3 atmospheres. Let x designate the depth we are looking for. We can thus write the equation

$$(1.001)^{\frac{x}{8}} = 3$$

whence, with the aid of logarithms, we learn that x is 8.9 km.

Hence, anyone of us could safely descend, without risk to life, to the depth of almost 9 km. Were the Pacific suddenly to dry up, we would be able to live practically everywhere on its bed.

In a Deep Mine

Who has got nearest to the centre of the earth—not, of course, in the imagination of a science-fiction novelist but in real fact? Miners, of course. We know (see Chapter 4) that the world's deepest mine is in South Africa. It is more than 3 km deep—that's the depth to which people have gone, because drills have bored down to even 7.5 km. This is what the French writer Dr. Luque Durten had to say after visiting a mine of Morro Veljo (about 2,300 m deep).

"The famous gold mines of Morro Veljo are some four hundred kilometres away from Rio de Janeiro. After a sixteen-hour train ride across rocky terrain, you descend into a deep valley surrounded on all sides by jungle. Here a British firm is mining gold at a depth to which man never descended before.

"The gold seam slants down and the mine follows it in six steps. The vertical shafts resemble wells and the horizontal workings—tunnels. It is extremely characteristic of modern society that the deepest mine dug in the earth's crust, man's boldest attempt to probe into the bowels of the earth, was undertaken in search of gold.

"Don canvas overalls and a leather jacket. Keep your eyes well open, the tiniest pebble dropping into the well may hit you hard. One of the mine captains will accompany you. You enter the first well-lit tunnel to shiver in the icy wind of only four degrees above zero coming from the ventilation system used to cool the deeper workings.

"You ride down the first seven-hundred-metre shaft in a cramped metal cage and find yourself in a second tunnel. You then go down the next shaft. Meanwhile the air grows warmer: You are already below sea level.

"In the next shaft the air burns your face. Perspiring freely and crouching beneath the low vaulted ceiling, you make for the roar of the drilling machines. Here miners, stripped to the waist and sweating profusely, work in clouds of thick dust. The water bottle makes round after round, endlessly. Don't touch the chunks of ore just hewn off. They run a temperature of 57°.

"What is the net result of this abominable and abhorrent reality? Some ten kilogrammes of gold a day."

In describing the actual physical conditions at the bottom of the mine and the intensified exploitation of the minehands, the French writer notes the high temperature but says nothing about the pressure. Let us then try to find out how much it would be at the depth of 2,300 m. Were the temperature there the same as at ground level, then, according to the familiar formula, the air density should increase by

$$(1.001)^{\frac{2,300}{8}} = 1.33 \text{ times.}$$

Actually, the temperature does not remain constant; it increases. Consequently, the air density does not rise so significantly. In the long run, the air at the bottom of the mine differs in density from that at ground level as the density of the air on a blazing hot summer day does from that of the icy air of winter time. That explains why visitors to the mine fail to notice anything.

165

Of much greater importance, on the other hand, is the considerable humidity of the air in such deep mines, which at high temperatures renders conditions intolerable. At a mine near Johannesburg, in South Africa (2,553 m deep), 100 per cent humidity is registered at 50° of heat. To combat this, air conditioning has been introduced; the cooling effect of the installation is equivalent to the presence of 2,000 tons of ice.

In a Stratosphere Balloon

So far we have been travelling—in the mind's eye, of course—into the depths of the earth in which we were assisted by a formula showing the dependency of pressure on depth. Let us now ascend and, using the same formula, see how atmospheric pressure changes at great altitudes. Naturally, we shall have to rewrite the formula which will now read:

$$p = 0.999^{\frac{h}{8}},$$

where p is the pressure in atmospheres, and h the altitude in metres. The decimal fraction of 0.999 has replaced our previous figure of 1.001, because with every 8 metres pressure, instead of increasing by 0.001, will, on the contrary, *decrease* by 0.001.

Now, firstly, how high up must we go to halve atmospheric pressure?

In this case p will be 0.5 and we shall be finding the altitude h. Our equation will be

$$0.5 = 0.999^{\frac{h}{8}},$$

which those of you familiar with logarithms will easily be able to solve. The answer is $h=5.6$ km, the height at which atmospheric pressure should be half the normal.

Let us now ascend still higher and follow in the wake of the intrepid Soviet aeronauts who reached the altitude of 19 and 22 km respectively, heights already in what is called the stratosphere. That is why balloons used for such ascents are known as stratosphere balloons. It was the Soviet stratosphere balloons *USSR* and *Osoviakhim-I* which in 1933 and 1934 set world altitude-records of 19 and 22 km respectively.

Let us now see what the air pressure would be like at these great heights. At 19 km it should be $0.999^{\frac{19,000}{8}} = 0.095$ atm = 72 mm, and at 22 km, $0.999^{\frac{22,000}{8}} = 0.066$ atm = 50 mm, of the mercury column.

However, according to the records other pressures were registered, to wit 50 mm at 19 km and 45 mm at 22 km. Where did we go astray?

Mariotte's law is quite valid for gases at such small pressures. Our omission was that we considered the air temperature constant throughout. Actually it drops noticeably with height. It is held that on the average the temperature drops by 6.5°C with every kilometre. This continues up to 11 km, beyond which the temperature remains constant at 56°C below zero for quite a considerable stretch further up. Taking this circumstance into account—which we cannot do with the means of elementary mathematics at our disposal— we would get results far closer to the real thing. For the same reason must you treat the results we obtained when gauging atmospheric pressure in deep mines as also approximate.

Chapter 7 *Heat*

Fans

When ladies fan themselves, they naturally feel refreshed. One might think that this occupation would be absolutely harmless for all others present, that they must be only grateful to the ladies for cooling the air. Let us see whether this is really so.

Why do we feel cooler when we fan ourselves? The air in direct contact with our face warms up. It is this warm invisible mask of air that "heats" the face, or in other words, prevents it from shedding any more heat. When the air around is still, this warm mask is but very slowly pushed up by the cool heavier air. When we fan away this warm mask, our face comes into contact with more and more new portions of non-warmed air, to which it sheds its warmth. That is how we cool ourselves.

Consequently, when fanning themselves, the ladies continually whisk the warmed mask of air, replacing it with nonwarmed air. This portion of air warms up to be again whisked away and replaced by still another portion of nonwarmed air.

Fanning thus accelerates air mixing and helps to

quickly equalise the temperature throughout the room. In other words, it refreshes the possessor of the fan at the expense of the cooler air enveloping the other people present. Then there is one more circumstance of significance in fanning, which I shall now proceed to tell you about.

Why Winds Make Us Feel the Cold More

You all know, of course, that in calm weather frosts don't bite so cruelly as they do in windy weather, but I suppose, not all of you clearly realise why. It is only the *living beings* that feel the cold more in a wind. It does not cause the thermometer to drop. The reason you feel the cold so keenly on a windy frosty day is, first of all, because the wind takes far more warmth away from the face—and from the body generally—than in calm weather, when the enveloping layer of air warmed up by the body is not so rapidly ousted by a new portion of cold air. The stronger the wind, the greater the mass of air that comes into contact with your skin every minute and, consequently, the greater the amount of warmth taken away from your body every minute. This alone is already enough to make you feel the cold.

But there is also another reason. Our skin always gives off moisture, even in cold air. To perspire we must have warmth. This warmth is derived from our body and from its enveloping layer of air. When the air is still, perspiration is slow since the layer of air adjacent to the skin is soon saturated with vapour—and in damp air, evaporation is not so intensive. But when the surrounding air is in motion and more and more new portions of it come into contact with the skin, perspiration is more profuse, requiring plenty of heat, which is taken away from the body.

How great is the wind's cooling effect? It depends on the wind's velocity and the air temperature. Generally speaking, it is much more significant than usually thought. Here is an illustration to show how greatly a wind reduces the body's skin temperature. Suppose the temperature of the air is 4°C above zero and that, for the time being, there is no wind. Then the body's skin temperature is 31°C. A light breeze of 2 m/sec, in which flags hardly flutter and leaves don't rustle at all, will cool the skin by 7°. A wind causing a flag to flutter, that is one having a speed of 6 m/sec, would already cool the skin by 22° right down to just 9° above zero.

Consequently, to gauge just how much a frost will affect us, it is not enough to go by the temperature alone; we must take wind velocity in account as well. The same degree of frost is harder to bear in Leningrad than in Moscow, because the average wind velocity on Baltic shores is 6 m/sec, while in Moscow it is only 4.5 m/sec. It is still easier to stand a frost near Lake Baikal, where the average wind velocity is only 1.3 m/sec. That is why the famous East Siberian frosts are far from being so harsh as we, being accustomed to the relatively strong winds of Europe, may think. On the contrary, East Siberia is distinguished by its almost complete absence of windy weather, especially in winter time.

The Desert's Scorching Breath

After what I have just said you may argue that on a blazing hot day a wind should refresh us. Why then do travellers speak of the *desert's scorching breath*? The contradiction is due to the fact that in tropical climes the air is usually warmer than the body. No wonder that these winds make it still hotter for people. This is a case when the air, instead of robbing the body of its warmth, on the contrary warms it up. Therefore the greater the mass of air that comes into contact with the body every minute, the more keenly we feel the heat, and though the wind increases the intensity of evaporation, this fails to help. That explains why desert folk wear warm robes and fur hats.

Do Veils Warm?

This is another problem for everyday physics. Ladies claim that veils give them warmth and that without them they would feel cold. Upon examining these flimsy affairs with their often rather big mesh, some gentlemen believe the claim the ladies make is but a trick of their imagination.

I think you, now that you have read my previous explanation, will put more stock in such claims. However big the mesh, air will penetrate a veil more slowly. The layer of air directly around the face warms up and the veil prevents the wind from whisking this warm mask away. Therefore when the ladies tell you that the veil keeps the face warm when out strolling on a breezy day with the temperature just a few degrees below zero—you can believe them.

Coolers

You have most likely seen, heard or read of these unglazed earthenware vessels with their curious property of being able to cool whatever they hold. They are common with southern nations being known as the "alcarraza" in Spain and the "goula" in Egypt, etc.

Their secret is very simple. As it seeps through the earthenware walls, the liquid slowly evaporates robbing both the vessel and its liquid contents of their heat ("latent heat of vaporisation").

However, travellers in the southern countries wrongly claim that these vessels cool their contents considerably. This effect depends on a number of factors. The hotter the surrounding air, the faster and more profuse the evaporation of the liquid moistening the outer wall of the vessel and, consequently, the more greatly is the liquid in the jar cooled. Humidity likewise plays a definite role. Should it be great, evaporation will be slow and the vessel will hardly cool its contents at all. On the contrary, in dry air evaporation is very intensive, and the cooling effect is greater. A wind will also accelerate evaporation and thus enhance the cooling effect—a fact you may have noticed when wearing a wet dress on a warm but windy day. The coolers can lower the temperature by not more than 5°. On a blazingly hot southern day, when the temperature is sometimes as much as 33°C, water in the cooling vessel will be 28°. As you see the cooling effect is practically nil, and it is not really for these purposes that these vessels are used. They are successfully used to keep *cold water* cold.

We might try to reckon the cooling effect of a cooler. Suppose this vessel can hold 5 litres of water and that, furthermore, 0.1 litre has evaporated. To evaporate 1 litre (1 kg) you would need on a hot day (33°C) 580 calories. With us it was 0.1 kg that evaporated; consequently, we needed 58 calories. If all this heat were taken from the water in the jar alone, its temperature would drop by 58/5 or by some 12°. However, most of the heat required for evaporation is taken from the walls of the jar itself and from the air around it. Furthermore, during the cooling process the water in the vessel is simultaneously warmed by the warm air around the jar. Hence, the actual reduction in temperature, that is, the cooling effect, reaches barely half our figure.

It is hard to say where the cooling effect is greater —in the sun or in the shade. The sun's heat accelerates evaporation but also increases the influx of warmth. I think the best place for a cooler is a draughty place in the shade.

"Icebox" Without Ice

A food cooler or an "icebox" without ice, is also based on the cooling effect that evaporation produces. It is a very simple affair: a wooden box—one of zinc-plated iron would be still better—with shelves inside for the food. On top place a tall vessel with cold water. Then dip in the tail end of a piece of canvas draped down the back of the box with its other end dipped in a vessel below the bottom shelf. The canvas soaks in the water from the vessel on top; meanwhile the water slowly evaporates as it moves through the canvas as through a wick, cooling all the compartments of the "icebox". This contraption should naturally be placed in the coolest spot in the room and every evening you should change the water so that it grow still colder during the night. It goes without saying, that the two vessels and the piece of canvas must be spotlessly clean.

The Greatest Heat One Can Bear

Man can stand heat much better than is usually thought. In southern latitudes he can bear up under a temperature noticeably above what we in moderate zones will scarcely tolerate. In Central Australia the summer temperature often rises to 46°C in the shade. There have even been cases when the temperature has jumped to 55°C in the shade. In the cabins of ships going through the Red Sea to the Persian Gulf the temperature has risen to 50° and more, despite continual ventilation.

The highest natural temperature ever observed in the world has been not more than 57°C. It has been registered in the "Valley of Death" in California. In Central Asia, the hottest place in the USSR, the temperature has never risen to more than 50°C.

You have probably guessed that the above-mentioned temperatures are all, of course, in the shade. Let me tell you why. The point is that it is only in the shade that a thermometer will register the right temperature of the *air*. After all if it were exposed to the sun, it might be heated up to a much greater temperature than that of the surrounding air. In short, there is no point in referring to readings of a thermometer exposed to the sun, when speaking of heat waves.

Experiments have been staged to determine the highest temperature the human body could bear. It has been found that when we warm up gradually in *dry air*, we can bear a temperature even above boiling point (100°C), as much as 160°C, as was demonstrated by the British physicists Blagden and Centry who for the sake of experiment spent hours on end in a heated bakery furnace. "You boil eggs and fry a steak in the air of a place where people could stay without doing any harm to themselves," Tyndall has noted in this connection.

Where does the explanation lie? In that our body actually repels this temperature, keeping down close to normal. It resists the heat by abundantly exuding sweat. This sweat absorbs much of the warmth from the layer of air directly enveloping the body, thus adequately lowering the temperature. The only essentials that need to be observed are for the body not to come into direct contact with the source of heat and for the air itself to be absolutely dry.

In Central Asia it is far easier to stand a heat of 37°C than a 24°C heat-wave in Leningrad. This of course is because of the very humid air in Leningrad and of the very dry air in Central Asia where rains are very few and far between.

Thermometer or Barometer?

As legend has it a certain Simple Simon did not dare to take a bath because, as he explained it, he had "stuck a barometer in the bath and it pointed to stormy weather".

Don't think it is always so easy, however, to distinguish between barometers and thermometers. There are types of thermometers or rather thermoscopes which we would have every justification for calling barometers and vice versa. An apposite illustration is afforded by the thermoscope that Heron of Alexandria invented (Fig. 83). In the sunshine the air in the up-

Fig. 83

Heron's thermoscope.

per part of the retort expands to press on the water and forces it to flow along the bent pipe, from the end of which it drips through a funnel to collect in the box below. In cold weather, on the contrary, air resilience in the sphere diminishes and the water in the bottom box is forced out by external air pressure up the straight pipe back into the sphere.

This instrument, however, reacts to varying barometric pressure; when external atmospheric pressure drops, the inside air, retaining its previous, higher pressure, expands, thus forcing part of the water up through the bent pipe and into the funnel. On the other hand, when external atmospheric pressure increases, part of the water from the bottom box is thereby forced into the retort. Each degree of difference in temperature will produce a similar difference in the volume of air inside the sphere (760 : 273 which is about a 2.5 mm difference in the height of the barometric column of mercury). In Moscow barometric fluctuations register 20 and more millimetres. This corresponds to 8°C on Heron's thermoscope—which means that this fall in atmospheric pressure may easily be taken for an 8° rise in temperature.

As you see this ancient thermoscope could well serve as a baroscope. We once had on sale water barometers which could have also served as thermometers, however this was something that neither the buyer nor even the inventor ever suspected.

What Is the Lamp Glass for?

Few know the long road that the lamp glass had to take before it reached its present shape. For thousands of years people used a flame without any glass guard for purposes of lighting. It needed the genius of Leonardo da Vinci (1452-1519) to introduce this major development. Leonardo da Vinci, however, used metal instead of glass and another three centuries passed before the transparent glass cylinder was introduced. As

176

you see, the lamp glass is the product of the ingenuity of dozens of generations.

What is it used for? I doubt whether any one of you would be able to provide the correct answer. The role of guarding the flame from the wind is but of secondary importance. The main purpose is to enhance the flame's *brilliance*, to accelerate the process of combustion. In other words, it acts as a chimney: it brings more air to the flame and thereby increases the draught.

Let us analyse this. The flame warms up the air column inside the glass much faster than the air around the lamp. Warming up and thus growing lighter, the air is pushed upwards—in conformity with Archimedes principle—by the heavier cooler air which enters from the bottom through the holes in the burner. We thus have a constant upward flow of air, a draft which takes away the products of combustion and brings in fresh air all the time. The taller the glass, the greater the difference in weight between the heated and the unheated air, and the more vigorous hence the influx of fresh air and, consequently, the process of combustion. This explains incidentally why factory chimneys are so tall.

It is curious to note that Leonardo da Vinci had a very clear notion of these things. In his manuscripts we find the following remark: "Wherever fire appears, an air current is induced around it: it is this air current that feeds and adds to the fire".

Why Does a Flame Never Extinguish Itself?

As soon as we stop to analyse the process of combustion we involuntarily ask ourselves: why does the flame never go out of its own accord? After all, combustion produces carbon dioxide and water vapour—two *noncombustibles*, which naturally cannot maintain combustion. Consequently, the moment it starts burning, a flame ought to be enclosed by noncombustibles, which in turn bar the intake of air. We know that without air combustion cannot continue and, consequently, the flame ought to go out.

But why doesn't this happen? Why does combustion go on until all the fuel is burnt out? Only because gases expand when heated and so become lighter. That is the only reason why the heated-up products of combustion do not remain where produced, that is, right in the flame's neighbourhood, but are immediately chased upwards by the clean incoming air. If the rule of Archimedes were not valid for gases—or if there were no such thing as gravity—every flame, after burning for a little while, would go out of its own accord.

It is easy to see how fatal the products of combustion are for a flame. You often invoke this point—without even suspecting it—to snuff out the flame. How do you put out a paraffin lamp for instance? You blow in from the top, in other words, you chase down towards the flame the noncombustible products of its combustion: as a result the flame dies as there is no more fresh air.

The Chapter Jules Verne Didn't Write

Jules Verne related in great detail the adventures of three brave travellers inside the moon-bound projectile. But he forgot to tell us how Michel Ardan did the cooking in these very unusual conditions. Most likely he didn't think cooking in space of any interest. So much the worse for the novelist. The thing is that inside the projectile, racing through space, *everything becomes weightless* (see Book One of *Physics for En-*

178

tertainment for detailed explanations of this highly interesting point). Jules Verne, regrettably enough, gave this the go-by, for, after all, one must agree that cooking in a kitchen that weighs nothing presents a science-fiction novelist with plenty of scope for imagination. Let me then try to make up to the best of my ability for what the talented author of *Journey to the Moon* missed. And don't forget—when you read my humble effort to simulate Jules Verne—that in the projectile there is no *gravity*, that not a single thing *weighs even a fraction of an ounce.*

Breakfast in Weightlessness

"We have not breakfasted as yet, my friends," Michel Ardan remarked. "And though we've lost our weight, I don't suppose we've lost our appetite. So, my friends, I'll now cook for you a weightless breakfast which will, I'm certain, offer the lightest dishes ever prepared."

And without waiting for a reply, the Frenchman began his gastronomic ministrations.

"Our water bottle is pretending to be empty," Ardan muttered to himself, as he endeavoured to uncork the big bottle. "But you won't fool me. I know why you're so light. There, I've got the cork out. Now go ahead and pour your weightless content into the pot!"

He tilted the bottle this way and that but the water wouldn't pour out.

"You're toiling in vain, my dear Ardan," said Nicoll, as he came to his help. "You must realise that in our projectile, where we have no gravity at all, water will never pour out! You must *shake* it out as if it were condensed syrup."

Ardan at once clapped the bottom of the tilted bottle. How surprised he was when a ball of water, the size of a fist, flew out of the bottle's neck.

"What's happened to the water?" exclaimed Ardan in astonishment. "I didn't expect this! Please tell me, my learned friends, what has happened?"

"That is merely a *drop* of water, my dear Ardan. In a weightless world with no gravity drops of liquid will assume any size. After all, it is thanks to gravity that liquids assume the form of the vessels that hold them, spill out in a stream, and so on. Since we are weightless, the liquid is left to its own inner molecular forces and so quite naturally assumes the form of a sphere, as the oil did in Plateau's famous experiment."

"I don't care a bit for your Plateau and his experiments! I must boil some water for the consommé and no molecular forces, I swear, will stop me!" the Frenchman declaimed in a temper.

He assailed the bottle furiously, endeavouring to shake the water out into the pot which hovered in the air, but everything seemed to be against him. The big drops of water crept about the pot as soon as they came into contact with it: they slid over onto the outside and soon the pot was enveloped in a thick layer of water. To boil the water in this state was out of the question.

"There is an illuminating experiment to demonstrate the great force of cohesion," the imperturbable Nicoll calmly told the enraged Frenchman. "Don't get so excited. You are dealing with an ordinary case of the wetting of solid bodies, only in this instance gravity is not interfering, and so we can see the entire process."

"It's a great pity that it isn't interfering!" Ardan heatedly objected. "And whether it's wetting or something else, I must have the water *inside* and not

180

arouna the pot. Look at it! No chef would ever consent to prepare consommé in these conditions!"

"You can easily stop that, if it's in your way," Mr. Barbicane inserted in a placatory tone. "Remember that water does not wet bodies covered with a thin film of grease. Grease your pot on the outside and you'll be able to keep the water inside."

"Well, that's what I call real scientific learning," the overjoyed Ardan exclaimed, as he took this advice. Then he lit the gas burner, wanting to put the pot on to boil, but again everything seemed to be against him. Now it was the gas burner that was acting capriciously. Its flame flickered for half a minute and went out. Ardan was bewildered. He nursed the flame but all his efforts were to no avail. The flame wouldn't burn.

"Barbicane! Nicoll! Isn't there a way of making this obstinate flame burn as it should, according to the laws of your physics and the gas company's regulations?" the dejected Frenchman appealed to his friends.

"There's nothing at all unusual or unexpected about it," Nicoll explained. "The flame is really burning as it should according to the laws of physics. As for the gas company's regulations, I suppose the firm would go to the dogs were there no gravity. As you know, combustion produces carbon dioxide and water vapour —gases that don't burn. Ordinarily these products of combustion don't keep near the flame because since they are warm and are consequently lighter, the fresh incoming air replaces them. But as we have no gravity here the products of combustion stay where produced. They envelop the flame and shut out fresh air. That is why the flame is so pale and goes out so quickly. Fire extinguishers by-the-by work precisely in this manner, surrounding the flame with a noncombustible gas."

"That means," the Frenchman interjected, "that if Mother Earth had no gravity, there would be no need for fire brigades, and fires would go out of their own accord, suffocated by their own breath? Is that right?"

"Quite. Meanwhile, to help prepare the consommé, light the burner once again and let's blow on the flame. In this way I think we can induce an artificial draught and make the flame burn as it would back home."

This was done. Ardan lit the burner once again and began cooking, meanwhile watching, not without a pinch of malice, how Nicoll and Barbicane alternately blew and fanned the flame to keep it burning. Deep down in his heart, the Frenchman thought that his friends and their science were wholly to blame for all this trouble.

"Ha-ha! You're very good as chimneys, I must say," Ardan vivaciously remarked. "I feel very sorry for you, my learned friends, but if you want to have breakfast hot you'll have to obey the laws of your physics."

A quarter of an hour passed. Then half an hour and then an hour. The pot showed no signs of boiling.

"You'll have to be patient, my dear Ardan. Ordinary water that has weight heats up rapidly. Why? Only because its different layers mix. The heated and, consequently, lighter lower layers are pushed up by the colder heavier upper layers and as a result the whole of the liquid quickly grows hot. Have you ever tried to heat up water from the top? Then the different layers won't mix because the heated layers remain still. Water is a very bad conductor of heat, as you know. Its conductivity is actually negligible. You can cause water to boil at the top, while having lumps of ice at the bottom. But here, in conditions of weightlessness, it makes no difference from which end we warm up the water. The different layers of water in the pot won't mix and the water should warm up very slowly. If you want to make it warm up faster, you must keep on stirring it."

Nicoll warned Ardan not to bring the water up to boiling point but to keep it a bit below that. At boiling point, he explained, there would be much steam. As in conditions of weightlessness, it would have a specific gravity equal to that of water—both would be nil—it would mix with the water and form a homogeneous foam.

Ardan was most annoyed when he untied the bag with the peas. He shook it slightly and the peas tumbled out into the air in all directions, bouncing on and off the walls. They nearly caused a great misfortune. Nicoll accidentally inhaled one of them and was almost suffocated. To ward off the danger and rid themselves of the perilous peas, our friends assiduously began to trap them with a butterfly net that Ardan had providentially taken along to "collect lunar butterflies".

Cooking in these conditions presented quite a formidable task. Ardan rightly remarked that in this contingency even a real chef would have thrown up the sponge. He also had quite an ordeal when he set about frying the steak. He had to clamp the meat down with a fork, because the resilient vapours of the frying oil beneath it kept on causing the half-done meat to bounce "up"—if one could use a word like that in a place where one has neither "up" nor "down".

The process of eating in this weightless world also presented an extremely queer spectacle. They hung suspended in mid air in diverse poses—not devoid incidentally of a certain note of extravagance—and kept on bumping their heads. Sitting was naturally out of the question. Chairs, couches, benches and the like are absolutely useless in a weightless world. Actually there would have been no need for a table, had not Ardan insisted upon having a real "breakfast table".

It was hard to cook the consommé but still harder to eat it. To begin with Ardan couldn't pour the weightless liquid out. The labours of a whole morning were almost lost, when, forgetting that the soup was weightless, he vexedly thumped the bottom of the overturned pot to get the obstinate soup out. A huge ball-shaped drop flew out—the soup itself! Ardan had to display a juggler's skill to trap and get back in the pot the consommé he had prepared with such difficulty.

Spoons failed to help: the soup wettened the spoon up to the finger-tips drooping from it like a solid veil. The three friends then greased their spoons with butter to prevent wettening, but that didn't help either. The soup assumed the form of a little ball and they couldn't get this weightless pill into their mouths.

Finally, Nicoll found a way out. He rolled some waxed paper into tubes and the three travellers sucked the soup up through them. They used the same method to drink water, wine, and all other liquids. (Many who read a previous edition of this book wrote to me wondering how one could drink in a weightless world even with the aid of the method suggested. After all the air in the projectile weighed nothing and, consequently, could not exert any pressure, which should thus make drinking by the sucking in of a liquid impossible.

Queerly enough, this view had some currency in the press. But it is quite obvious that in these conditions the weightlessness of air has no effect on pressure. Air exerts pressure in a closed space, not at all because it has weight but because, as a gaseous body, it tries to expand without end. In *unclosed* spaces on our planet it is *gravity* that is the barrier to expansion and it was this customary interrelation that misled my critics.)

Why Does Water Put out Fire?

Though a simple question not all supply the right answer. I hope you won't take it amiss if I stop to explain briefly what water actually does to fire. Firstly, as soon as it comes into contact with the burning object, water turns into steam, in which process it deprives the burning object of much of its heat. After all, to transform boiling water into steam we need five odd times more heat than is required to heat the same amount of cold water to boiling point. Secondly, the steam thus produced occupies a space hundreds of times bigger in volume than the water giving rise to it. The steam envelops the burning object and keeps fresh air away. Without air combustion is impossible.

To cause water to act as a still better extinguisher of fires, gunpowder is sometimes added to it. There is logic in this paradox. Gunpowder burns out quickly, giving off in the process a large quantity of noncombustible gas. This gas envelops the burning object and complicates combustion.

Fighting Fire with Fire

You most likely know that the best and sometimes only way of fighting a forest or prairie fire is to set fire to the forest or prairie from the other side. The second fire moves towards the first, and, by destroying combustible material, deprives it of fuel. As soon as

Fig. 84

Fighting fire with fire.

they meet, the two walls of fire die, devouring each other, as it were.

Most likely many of you have read of this in Fenimore Cooper's *Prairie*. Surely, you haven't forgotten that tense and dramatic moment of suspense when the old trapper saves the travellers from a fiery death? Here is the extract.

"...the old man ... suddenly assumed a decided air. ...

"'It is time to be acting,' he said ...

"'You have come to your recollections too late, miserable old man,' cried Middleton; 'the flames are within a quarter of a mile of us, and the wind is bringing them down in this quarter with dreadful rapidity.'

"'Anon! the flames! I care but little for the flames... Come, lads, come. ... Put hands upon this short and withered grass where me stand, and lay bare the 'arth'... A very few moments sufficed to lay bare a spot of some twenty feet in diameter. Into one edge of this little area the trapper brought the females, directing Middleton and Paul to cover their light and inflammable dresses with the blankets of the party. So soon as this precaution was observed, the old man approached the opposite margin of the grass, which still environed them in a tall and dangerous circle, and selecting a handful of the driest of the herbage, he placed it over the pan of his rifle. The light combustible kindled at the flash. Then he placed the little flame in a bed of the standing fog, and withdrawing from the spot to the center of the ring, he patiently awaited the result.

"The subtle element seized with avidity upon its new fuel, and in a moment forked flames were gliding among the grass. ...

"'Now,' said the old man, holding up a finger, and laughing in his peculiarly silent manner, 'you shall see fire fight fire. ...'

"'But is this not fatal?' cried the amazed Middleton; 'are you not bringing the enemy nearer to us instead of avoiding it?'... As the fire gained strength and heat, it began to spread on three sides, dying of itself on the fourth, for want of aliment. As it increased and the sullen roaring announced its power, it cleared everything before it, leaving the black and smoking soil far more naked than if the scythe had swept the place. The situation of the fugitives would have still been hazardous had not the area enlarged as the flame encircled them. But by advancing to the spot where the trapper had kindled the grass, they avoided the heat, and in a very few moments the flames began to recede in every quarter, leaving them enveloped in a cloud of smoke, but perfectly safe from the torrent of fire that was still furiously rolling onward.

187 "The spectators regarded the simple expedient of

the trapper with that species of wonder, with which the courtiers of Ferdinand are said to have viewed the manner in which Columbus made his egg stand on its end. ..."

Incidentally this method of fighting forest and prairie fires is not so simple as it may seem at first glance. Only a very skilled hand at it may use it. A novice would only make matters worse.

You will understand what I am driving at if you ask yourself: why did the fire, which the trapper lit, run towards the other fire and not contrariwise? After all the wind was blowing into the faces of the travellers and was driving the flames towards them. Shouldn't the fire that the trapper started have gone the other way? In that case the travellers would have found themselves hemmed in by a circle of flames and would have certainly perished.

So what was the trapper's secret? In that he knew a simple law of physics. Though the wind was blowing from the burning prairie into the faces of the travellers, in front of them, right near the flames, a reverse air current was blowing towards the fire. Indeed, warmed up by the fire below, the air above it grows lighter and is pushed up by cooler fresh air flowing in from the prairie. This explains why near the fire's fringes the draught is directed towards it.

The fire-fighting fire must be started when the original fire has drawn close enough for one to feel this draught. That is why the trapper didn't hurry, calmly waiting for the necessary moment. Had he fired the grass too early, before the counter-draught had set in, his fire would have spread in the opposite direction and would have placed the travellers in a hopeless predicament. Too late would be just as fatal because then the fire would be too close.

Can We Boil Water in Boiling Water?

Take a small bottle or a jar, fill it with water and put it in a pot full of water placed on fire, but so that it does not touch the bottom. To abide by the last condition, you will have to suspend it in a wire loop. One would think that when the water in the *pot* boils, the water in the jar should boil too. However, no matter how long you wait, that won't happen. The water in the jar will be very hot, but it won't boil. Boiling water, we find, is not hot enough to boil water.

This comes as quite a surprise, doesn't it? However, it should be expected. After all, to bring water up to boil, it is not enough to heat it up to 100°C. It needs more heat to transform water into its next state, steam.

Pure water boils at 100°C. In ordinary conditions it never rises above this temperature, however much we heat it. This means that the source of heat we are using to heat the water in the jar has a temperature of 100°C and no more and thus can heat the water in the jar also up to 100°C, and no more. As soon as the temperatures equalise, *the water in the pot can no longer impart any more heat to the water in the jar.*

To sum up: by heating water in the jar in this way we shall not be able to give it that extra amount of heat which is required to transform it into steam. (Each gramme of water heated up to 100°C requires another 500 odd calories to turn into steam.) That is why the water in the jar doesn't boil, though it is hot.

You might want to know what difference there is between the water in the jar and the water in the pot. After all the water is the same in both vessels: the only difference is that the water in the jar is separated from the water in the pot by a wall of glass. Then why isn't it affected in the same way as the water in the pot?

Precisely because this wall of glass prevents the water in the jar from participating in the currents that mix all the water in the pot. Every particle of water in the pot will come into direct contact with the pot's bottom. Meanwhile the water in the jar will come into contact only with the boiling water in the pot.

So, as we have noticed, it is impossible to boil water in pure boiling water. But as soon as you add some salt to it, the picture changes; salt water boils not at 100°C but at a somewhat higher temperature, and, consequently, can bring the pure water in the jar to boil.

Can We Boil Water in Snow?

Well, you will say, if boiling water can't do the trick, how dare we speak of snow? Don't jump to conclusions, though. Do the following experiment first, using that same glass jar which we employed in the previous experiment. Fill half of it with water and immerse it in boiling *salty water*. As soon as the

Fig. 85

Water boiling in a retort after cold water has been poured over it.

water in the jar boils, take it out and quickly cork it tightly. Now turn it over and wait for the boiling to stop. Then pour a little boiling water over it. The water inside won't boil. But you need only put a little snow on the bottom of the jar or even pour a little cold water over it—as is shown in Fig. 85—for

Fig. 86

What happens to a tin can when unexpectedly cooled.

the water in the jar to start boiling at once. Snow has done what boiling water failed to do!

This is all the more mysterious because when you touch the jar with a finger, you will not find it very hot. Still you will see the water inside boiling. The answer lies in the fact that the snow cools the walls of the jar. The steam inside condenses into drops of water. But since the air in the jar was pushed out when the water boiled, now the water in it is subjected to a much smaller pressure. You already know that at a lower pressure liquids boil at a lower temperature. Consequently, we have boiling water in the jar but boiling water that isn't hot.

191

If the walls of the jar are very thin, the sudden condensation of the steam inside may produce something like a minor explosion. Failing to encounter an adequate resistance from inside the jar, the pressure exerted by the outer air may crush it. (Incidentally, the word "explosion" is not a happy term or phrase in this case.) It is therefore better to take a spherically shaped jar—a retort, for instance, so that the outer air press on its arched sections.

It is safest of all to stage this experiment with a tin can. After boiling some amount of water in it, screw the top on tight and pour cold water over it. The tin containing the steam will be crushed by the pressure of the outer air as this steam has condensed into water in the process of cooling. The tin will seem to have been struck at with a heavy mallet (Fig. 86).

"Barometer Soup"

In his *A Tramp Abroad* Mark Twain describes the following event—naturally a fictitious one—that happened during mountain climbing in the Alps.

"Our distresses being at an end, I now determined to rest the men in camp and give the scientific department of the Expedition a chance. First, I made a barometric observation, to get our altitude, but I could not perceive that there was any result. I knew, by my scientific reading, that either thermometers or barometers ought to be boiled, to make them accurate; I did not know which it was, so I boiled both. There was still no result; so I examined these instruments and discovered that they possessed radical blemishes: the barometer had no hand but the brass pointer and the ball of the thermometer was stuffed with tinfoil. ...

"I hunted up another barometer; it was new and perfect. I boiled it half an hour in a pot of bean soup which the cooks were making. The result was unexpected: the instrument was not affected at all, but there was such a strong barometer taste to the soup that the head cook, who was a most conscientious person, changed its name in the bill of fare. The dish was so greatly liked by all, that I ordered the cook to have barometer soup every day. It was believed that the barometer might eventually be injured, but I did not care for that. I had demonstrated to my satisfaction

Fig. 87

Mark Twain's researches.

that it could not tell how high a mountain was, therefore I had no real use for it."

But joking apart, let us try to answer the question as to what we should have really "boiled"—the thermometer or the barometer? The answer is: the thermometer. And here is why.

We have seen from previous experience that the lower the pressure exerted on water, the lower the temperature at which it boils. Since atmospheric pressure decreases the higher up you go, consequently, the lower should the temperature be at which water boils. Here is a table showing the temperatures at which pure water boils under different atmospheric pressures.

193

Temperature at which water boils, Centigrade	Barometric pressure, mm
101	787.7
100	760
98	707
96	657.5
94	611
92	567
90	525.5
88	487
86	450

At Berne in Switzerland, where the mean atmospheric pressure is 713 mm, water boils in an open-mouthed vessel already at 97.5°C, while on top of Mont Blanc, where the barometer registers a pressure of 424 mm, water boils at a temperature of only 84.5°C. With every kilometre up, the temperature at which water boils drops by 3°C. Consequently, if we measure the temperature at which water boils, or—to use Mark Twain's expression— "boil the thermometer," *by addressing ourselves to the appropriate table* we shall be able to find the altitude. To do that we must, of course, have the table that Mark Twain "simply" forgot.

The instruments used for this purpose—they are called hypsometers—are just as easy to carry as metal barometers, but give a far more accurate reading.

Of course, a barometer will also tell us how high up we are, as it doesn't have to be "boiled" to register atmospheric pressure and, after all, the higher up we go the lower pressure gets. In this case, we shall again need a table to show how atmospheric pressure drops the more we ascend from sea level or know, at least, the appropriate formula. The humorist, however, confused everything and so decided to "cook barometer soup."

Is Boiling Water Always Hot?

The brave batman Ben-Zouf, whom you might remember from Jules Verne's *Hector Servadac*, was firmly convinced that boiling water would always be scaldingly hot wherever it was boiled. I imagine he would have stuck to his guns to the end of his days had he not found himself with Servadac on a comet. This capricious celestial body collided with Mother Earth to slice off a chunk with our two heroes on it and carry it along with it on its elliptical orbit. That was when the batman found out for the first time in his life that boiling water was not identically hot everywhere. He made this discovery unexpectedly when preparing breakfast.

"Ben-Zouf filled the pot with water and put it on to boil. In his hands he held the eggs. They seemed to him to be empty inside, being as light as feathers.

"When in under two minutes the pot began to boil, Ben-Zouf exclaimed:

"'Faith, how hot the fire must be!'

"'It is not the fire that is hotter,' Servadac returned after brief thought, 'but the water that boils sooner.'

"He then took the Centigrade thermometer down from the wall and dipped it into the boiling water. It showed exactly sixty-six degrees.

"'God save us,' the captain cried, 'water is boiling at sixty-six degrees instead of one hundred!'

"'Well, captain?'

"'So let me advise you, Ben-Zouf, to boil the eggs for a quarter of an hour.'

"'They will be hard-boiled then.'

"'No, my friend, on the contrary, they will be just about right.'

"Evidently this was all because the height of the atmosphere had diminished. The column of air pressing on the ground had grown shorter by about a third, which explained why the water, being subjected to a smaller pressure, boiled at sixty-six degrees instead of one hundred. This would have also happened on a mountain peak eleven kilometres above sea level. If the captain had had a barometer on hand, it would have indicated this lessening of atmospheric pressure."

We shan't question their observations. They claim that water boiled at 66°C and we shall accept it at face value. But it is extremely doubtful whether the two men would have felt so fit in the rarefied atmosphere around them.

Jules Verne quite rightly notes that we would see water boiling at this temperature at the height of 11,000 m. At this altitude, according to calculations, water should really boil at 66°C (as we noted earlier, the temperature at which water boils drops by 3°C with every kilometre up and so to get water boil at a temperature of 66° one must ascend $34 : 3 \approx 11$ km). In this case atmospheric pressure should be only 190 mm of the mercury column which is exactly a quarter of normal atmospheric pressure. It is practically impossible to breathe air rarefied to such an extent. After all this altitude is already in the stratosphere. We know that pilots who reached this height without oxygen masks lost consciousness. Meanwhile Servadac and his batman felt more or less fit. It is a jolly good thing that Servadac did not have a barometer on hand; otherwise Jules Verne would have had to make it register not the figure which it should have shown according to the laws of physics.

Had our two heroes found themselves not on the imagined comet but on Mars, for instance, where atmospheric pressure is no more than 60-70 mm, they would have drunk still colder boiling water, heated up to only 45°C.

Conversely, very hot boiling water can be obtained at the bottom of deep mines where atmospheric pressure is much greater than at ground-level. At 300 m down water boils at 101°C, and at 600 m down already at 102°C.

When pressure is greatly increased, water boils even in the boiler of a steam engine. At 14 atm, for instance, water boils at a temperature of 200°C. On the contrary, under the bell jar of an air-pump one could get water to boil at ordinary room temperature, in which case the "boiling water" would be only 20°C hot.

Hot Ice

We have just been speaking of cool boiling water. But there is still a more surprising thing—*hot ice*. We have grown accustomed to the idea that water cannot exist in a solid state at a temperature above 0°C. However, the physicist Bridgman demonstrated that this is not at all so. Under very high pressures water solidifies and stays in that state at temperatures way above 0°C. Generally speaking, Bridgman proved that there could be more than one kind of ice.

The ice he called "Ice No. 5" is obtained under the monstrous pressure of 20,600 atm, and stays solid at a temperature of 76°C. It would scorch our finger-tips—provided, of course, that we would be able to touch it. We can't do that, however, because it is formed under a powerful press in a vessel with very thick walls made of the best grades of steel. We can't even see it and all that we know about its properties has been learned indirectly.

It is curious to note that this "hot ice" is denser than ordinary ice and even denser than water. It has a specific gravity of 1.05. It would sink in water— whereas ordinary ice, as you well know, floats.

Cold from Coal

The obtaining of cold, not heat, from coal is not at all a fantasy. It is effected daily at factories which make what is called "dry ice". Here the coal is burned in boiler-drums, the smoke it gives off is purified and the carbon dioxide it contains trapped in an alkaline solution. The pure carbon dioxide subsequently separated by heating is then cooled and compressed and liquefied under a pressure of 70 atmospheres. This is the selfsame liquid carbon dioxide that is shipped in thick-walled cylinders to factories making fizzy drinks or that is used for industrial purposes. It is cold enough to freeze ground—as was done when the Moscow Underground Railway was built. However, there are many cases in which we need solid carbon dioxide or what is known as dry ice.

Dry ice is derived from liquid carbon dioxide by fast vaporisation under a reduced pressure. Outwardly, chunks of dry ice sooner resemble pressed snow than ice and, generally speaking, differ greatly from solidified water. This ice is heavier than ordinary ice and sinks in water. Despite its extremely low temperature—78°C below zero—you won't feel the cold, provided you hold it gingerly, because the carbon dioxide gas which forms as soon as the piece comes into contact with our warm finger-tips, protects the skin from the cold. Only if you clasp a chunk of dry ice tight, will you run the risk of freezing your fingers.

The name "dry ice" is exceedingly apt as it emphasises its salient physical characteristic. It is, indeed, never wet and will never wetten anything it comes into contact with. Warmed, it immediately turns into gas, skipping the liquid state, because carbon dioxide cannot exist in a liquid state at a pressure of but one atmosphere.

This characteristic feature of dry ice, plus its low temperature, makes it an invaluable cooling agent for practical purposes. The products preserved with its aid never grow damp and are furthermore protected from spoilage—from mildew and mould—by the carbon dioxide gas that obstructs the development of micro-organisms. Neither will insects or rodents be able to live in such an atmosphere. Finally, carbon dioxide provides a reliable aid for fighting fire. A few lumps of dry ice thrown into burning petrol extinguish the flames at once. This has all added to its popularity in both industry and the household.

Chapter 8 *Magnetism and Electricity*

"Loving Stone"

Such is the poetic name the Chinese have bestowed on natural magnets. *Chu shi*, the "loving stone", the Chinese say, attracts iron as a tender mother will draw her children to her bosom. Curiously enough, the French, a people living at the other end of the Old World, have a similar name for the magnet. They call it *aimant* which means both *magnet* and *loving*.

The *loving* power of natural magnets is small and the Greek name for the magnet, "Hercules' stone", is, consequently, rather naive. If a natural magnet's moderate power of attraction had the ancient Hellenes wonderstruck, I wonder how amazed they would have been to see the modern magnets used at iron and steel plants, magnets able to lift huge chunks weighing several tons. True, these are not natural magnets, but electromagnets, or, in other words, hunks of iron magnetised by an electric current flowing through the coil wound round them. However, both cases

200

manifest one and the same sort of attraction—magnetism.

Don't think that a magnet attracts only iron and nothing else. There are some other metals which a powerful magnet will attract—though with not so great a force. These are nickel, cobalt, manganese, aluminium, gold, silver and platinum. The properties possessed by so-called diamagnetic bodies such as

Fig. 88

Candle flame between the two poles of an electric magnet.

zinc, lead, sulphur and bismuth, are still more remarkable. These metals are repelled by a powerful magnet.

A magnet can also attract, or repel, liquids and gases. True it must be very powerful in order to exert any influence. For example, a magnet can attract pure oxygen. If we were to fill a soap bubble with oxygen and place it between the holes of a powerful electromagnet, its invisible magnetic forces would distend the soap bubble between the two poles quite noticeably. Placed between the poles of a powerful magnet the flame of a candle will also change its habitual form and clearly exhibit a sensitivity to magnetic power (Fig. 88).

The Compass Problem

We have grown accustomed to thinking that a compass needle always points with one end North and the other South. Therefore, the following question might seem absurd: where in the world will a magnetic needle point North with both ends? My next question may seem just as absurd: where in the world will a compass needle point South with both ends?

I bet you'll claim that there can't be any places like that on our planet. However, they do exist. If you recollect that the earth's magnetic poles do not coincide with its geographical poles, you will have probably guessed what places I mean. Where will the needle of a compass point when at the South geographical pole? One end will point to the closest magnetic pole and the other to the opposite magnetic pole. But whichever way you go from the South geographical pole, you will always be going *North*— there is, indeed, no other way to go, except North. Consequently, there the compass needle will point North with both ends. Similarly will the needle of a compass at the North geographical pole point with both ends South.

Lines of Magnetic Forces

Fig. 89 shows a curious phenomenon reproduced from a photograph. It shows a great many nails sticking out like bristles from an arm resting on the poles of an electromagnet. The arm itself does not sense any magnetic attraction at all. Meanwhile the magnetic forces pass through it invisibly, causing the iron nails to obey them and dispose themselves in a regular pattern exhibiting the direction of the magnetic forces.

Since we have no magnetic organ of sense, we can

only guess at the existence of the magnetic forces emanating from a magnet. (It would not be devoid of interest to wonder what we would really feel if we did have a magnetic sensitivity. Kreidel managed to impart a sort of magnetic sensitivity to crawfish. He noticed that young crawfish poked tiny pebbles into their auditory organs, which tended to weigh on the sensitive hair comprising a component part of

Fig. 89

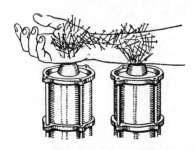

Magnetic forces in the arm.

their balancing organ. The human ear also has similar pebbles or stones which are called otoliths. They are located near the main auditory organ. Acting in the vertical direction, these stones point to the direction of the force of gravity. Kreidel was able unnoticeably to insert iron filings instead of these stones into the ears of the crawfish. When a magnet was brought up, the crawfish disposed itself in a plane perpendicular to the resultant of the magnetic force and force of gravity.

"Of late a modification of these experiments has been staged on a human being. Kehler glued minute iron filings to the ear drum, the result being that the ear received the oscillations of magnetic force as a sound." [Prof. O. Wiener.]) It is, however, simple enough to detect indirectly the pattern of the align-

Fig. 90

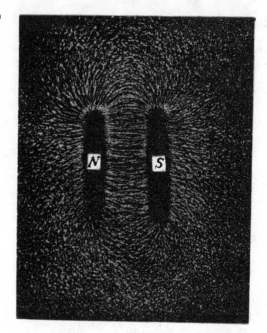

Pattern of iron filings on a piece of cardboard, under which a magnet has been placed. Reproduced from a photograph.

ment of these forces. This is best done by using small iron filings.

Take a piece of smooth cardboard or a glass plate and sprinkle filings over it in a thin layer. Then place the cardboard or glass with the filings on top of a bar magnet and gently flick the cardboard with a finger. Since the magnetic forces pass "freely through cardboard and glass, the iron filings will be magnetised. When jolted out of place by the flicking of the cardboard, the magnetic forces cause them to rearrange themselves, as a magnetic needle would at each particular point—lengthwise along the magnetic lines of force. As a result the filings dispose themselves in rows, graphically revealing the pattern of the invisible magnetic lines—as shown in Fig. 90. The magnetic forces produce an intricate system of curves, which strike out radially from each pole of the magnet and link up to form, now short, now long arcs between the two poles. The filings show what a physicist pictures in his mind and what invisibly surrounds every single magnet. The closer to the poles, the thicker and more distinct the lines are. Further away they grow hazier, well illustrating the weakening of magnetic forces of attraction with distance.

How Is Steel Magnetised?

To answer this often-asked question, we must first of all understand the difference between a magnet and a nonmagnetic bar of steel. Every atom of iron in a bar of steel, whether magnetised or not, may be imagined as a tiny magnet. In a nonmagnetic state these baby magnets are haphazardly orientated, with their poles completely neutralising each other (Fig. 91a). In a magnet, on the contrary, all the tiny magnetic units are disposed in a regular pattern in which like poles are set in the same direction—as is shown in Fig. 91b.

Now what happens in a bar of steel when magnetised? The magnet's attractive forces cause the magnetic particles in the steel bar to line up with their South or North poles pointing in one and the same direction. Fig. 91c provides a graphic apposite illustration. The magnetic units swing round at first with their south poles pointing to the magnet's north pole,

Fig. 91

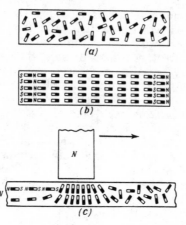

Disposition of the atom magnets in a nonmagnetised piece of steel (a); same in a magnetised piece of steel (b); the effect of the pole of a magnet on the atom magnets of magnetised steel (c).

then, when the magnet is moved further, they orientate themselves in the direction of its movement, their south poles turned inward.

This shows how one uses a magnet to magnetise steel. One must place one pole of the magnet on the end of a bar of steel and, pressing, move it to the other end. This is one of the simplest and oldest ways of magnetising steel, suitable, however, only for obtaining small and feeble magnets. Powerful magnets are obtained by drawing on the properties of electric currents.

Giant Electromagnets

At iron and steel mills you may see electromagnetic cranes lifting huge loads. They render an inestimable service, as they can hoist and transport, without any special attachment, huge chunks of iron or machine parts weighing dozens of tons. They can also carry—unpacked and uncrated—sheet iron, wire, nails, metal scrap and other materials, which it would take much time and energy to transport in any other way.

Figs. 92 and 93 show you how useful these lifting magnets are. How much trouble it would be to collect and carry the pile of iron plates which the powerful magnetic crane depicted in Fig. 92 lifts in one

Fig. 92

Electromagnetic crane transporting iron plates.

Fig. 93

Electromagnetic crane transporting barrels of nails.

go! We derive an advantage expressed not only in the saving of labour but also in the simpler way of doing things. Fig. 93 shows a magnetic crane carrying barreled nails, lifting six barrels at a time! At one iron and steel plant four magnetic cranes, each carrying ten rails at once, do the job of 200 workers.

I have already mentioned that you do not have to attach these loads to the crane, because while electric currents pass through the coil of the electromagnet, nothing will ever drop down. However, should the line break there is bound to be an accident, and there were such accidents when lifting electromagnets were first introduced. "At one American plant," we read in a technical journal, "an electromagnet was carrying iron ingots from the train to the furnace. Suddenly there was a mishap at the Niagara waterfalls power station supplying the electricity and the line went dead. The mass of metal broke away from the electromagnet and crushed to pulp a worker below. To prevent a repetition of such accidents and also save on electricity consumption special gadgets are supplied. As soon as the lifting magnet has snatched up the items to be carried, they are caught up from beneath by huge steel pincers and while the load is being shifted from place to place the electric current is switched off."

The electromagnets depicted in Figs. 92 and 93 are 1.5 m in diameter and each is able to lift up to 16 tons—as much as a loaded freight car. In a day's work one magnet alone can transport more than 600 tons. There are electromagnets which can lift in one go as much as 75 tons—a whole locomotive!

It might have struck you as very convenient to carry *hot* iron bars in this way. Unfortunately, iron is susceptible to a magnet's attractive force only up to a certain temperature. Red-hot iron loses its magnetic properties. A magnet heated up to the temperature of 800°C no longer possesses its magnetic properties.

Electromagnets are widely used at modern metalworking establishments to hold in place or move items of iron, steel, or pig iron, and hundreds of various devices have been evolved to simplify and speed up operations.

Sometimes circus magicians also resort to electro-magnets; one can well imagine how effective these tricks should be. Dary, the author of the well-known book *Electricity and Its Use*, recounts the following story which a French magician told him about a performance in Algeria. The trick made a tremendous impression upon the audience who knew nothing about electricity; they thought that they were seeing a real miracle.

"I had on the stage," the magician's story went, "a small iron-tipped box with a handle on top. I invited a strong man to step forth. In response an Arab of medium height but sturdily built, a sort of local Hercules, came up. He walked up cheerfully, grinning slyly the while, and stopped opposite me.

"'Are you strong?' I asked, looking at him from tip to toe.

"'Yes,' said he.

"'Are you sure you will always be strong?'

"'Certainly,' came the answer.

"'You're wrong,' I said. 'In the twinkling of an eye I can rob you of your strength and you will grow as weak as a little child.'

"The Arab gave a disbelieving grin.

"'Come here,' I said, 'and lift up this box.'

"The Arab bent over, lifted it up and then inquired:

"'Is that all?'

"'You just wait a moment,' said I. "Assuming a serious mien I made a commanding gesture and solemnly uttered:

"'You are now weaker than a woman. Lift that box once again.'

209

"Undaunted by my magic passes, the Arab again bent over and took hold of the box. However, this time it resisted. Despite his desperate efforts, it remained as if rooted to the spot. He strained himself might and main, exerting a force enough to lift an enormous load. But, it was all to no avail. Exhausted, puffing, and crimson with shame, he gave up. Now he believed in my magic."

The secret was simple enough. The iron-tipped box stood on a platform that was actually the pole of a powerful electromagnet.

It was easy enough to lift the box while the current was off, but as soon as the current flowed through the coil, even three strong men would have failed to lift it.

Magnet in Agriculture

Still more curious is the service that magnets render in agriculture. They help the farmer to separate the seeds of cultivated plants from the seeds of weeds.

Weeds have hairy seeds that cling to the wool of passing animals and thus spread to-long distances away from the mother plant. The farmer has availed himself of this characteristic feature which weeds have evolved in the aeons of their struggle for survival, to separate the hairy weed seeds by means of a magnet from the smooth seeds of such useful plants as flax, clover and alfalfa. The contaminated mixture of seeds is sprinkled with finely grated iron filings which adhere to the hairy seeds of weeds. As soon as they find themselves in the field of a sufficiently powerful electromagnet, the seeds separate, the magnet attracting all seeds to which the soft iron filings have adhered.

At the beginning of this book I had occasion to refer to Cyrano de Bergerac's amusing *History of Lunar and Solar States*. He happens to describe a curious flying machine based on magnetic attraction which one of his characters used to travel to the moon. Here is the extract:

"I ordered a light cart to be made of iron. As soon as I was comfortably seated, I began to toss up a magnetic ball. My iron cart at once ascended in its wake. Every time I reached the ball, I threw it up still further. The cart ascended even when I simply lifted the ball. After I had thrown the ball up time and again, the cart lifted me to a place from whence I began my descent on the moon. And since, at this moment, I held fast to the magnetic ball, the cart pressed closely to me. Not to break my neck on the descent, I threw the ball up to delay the cart's fall by its attraction. At a distance of some six or seven hundred yards from the moon's surface, I began throwing up the ball at right angles to the direction of my descent until the cart drew quite near. Then I jumped out to the sand below."

Everyone—including Cyrano de Bergerac himself—must have realised that his project was utterly unfeasible. But I am not sure all will be able to say why. Is it because you can't throw a magnet up while sitting in an iron cart? Or because the cart will not be attracted by the magnet? Or is it something else?

Let me tell you that you *could* throw a magnet up and it *would* attract such a cart if powerful enough. However, the flying machine would not move a single inch up.

Have you ever thrown a heavy article ashore from a boat? There is no doubt that, if observant enough, you felt the boat move away from the shore. This happened because while you were imparting an impetus to the object you were throwing in one direction, your muscles compelled your body and the boat with it to recoil. This is a manifestation of that selfsame law of the equality of action and reaction which we have already mentioned more than once. In the throwing up of the magnet you would have the same thing happening. As the man in the cart threw up the magnetic ball—which would cost him quite an effort, as the cart would also attract it—he would inevitably push the cart down. Coming together due to mutual attraction, the cart and the ball would merely regain their initial positions. Hence it is quite clear that even if the cart were as light as a feather, the throwing up of the magnetic ball would only cause it to swing up and down from a definite mean position. We would never get the cart to translate.

In the mid-17th century when Cyrano de Bergerac wrote his book, the law of action and reaction was still unknown. It is therefore doubtful whether the French satirist would have ever been able to state clearly why his amusing project was impracticable.

"Mahomet's Tomb"

One day, when an electromagnetic crane was in operation, a worker noted that the electromagnet had attracted a heavy iron ball attached to a short chain riveted to the floor. Since the chain prevented the ball from coming into direct contact with the magnet, leaving a gap, the width of a hand, in between, he saw the unusual sight of a ball and chain jutting vertically up. The magnet was so powerful that the chain remained in this position, even when the worker climbed it. (This, incidentally, indicates the electro-

Fig. 94

Iron chain with ball, standing upright

magnet's tremendous attractive power because a magnet will exert a far weaker force the greater the gap between its pole and the object being attracted. A horse-shoe magnet capable of holding a 100-gramme load in direct contact, loses half of its power of attraction when a sheet of paper is inserted between it and the load. That is why you will never find magnet ends painted, though paint offers protection from corrosion.) A photographer who chanced to be on hand did not let the opportunity slip and the picture he took—depicting a man suspended in mid air, somewhat like the legendary tomb of Mahomet—is reproduced in Fig. 94.

Incidentally, a few words about this tomb. Followers of Islam are convinced that the tomb containing the "Prophet's" remains rests on a cushion of air. Could this be possible? "It is said," Euler wrote in his *Letters on Sundry Physical Materials*, "that Mahomet's coffin is suspended in mid air by a magnet. This does not seem improbable because there are man-made magnets capable of lifting as much as 100 pounds." (This was written in 1774 at a time when electromagnets were unknown as yet.)

This explanation is untenable. Were such a method (*attraction of a magnet*) applied, the state of equilibrium could exist only for an instant, because the slightest tremor, even the slightest breath of air, would have been quite enough to upset the balance and the coffin would have either dropped to the floor or ascended to the ceiling. To all practical intents, it is just as impossible to retain it in a state of suspended immobility as it is to place a cone on its vertex—even though the latter is theoretically probable.

However, it is quite possible to simulate the "Mahomet's tomb" phenomenon with the help of a magnet, the difference being that we should have to draw on forces not of mutual *attraction* but of mutual *repulsion*. (Even people who have studied physics are prone to forget that magnets both attract and repel.) Like poles of magnets repel each other. Two magnetised bars of iron, arranged so that their like poles point to each other, will repel each other. If we chose a magnet of the proper weight, we can easily get it to hover above a second magnet in a state of steady equilibrium without touching the latter. However, we shall have to use nonmagnetic props, made of glass, for instance, to prevent the hovering magnet from swinging into a horizontal plane. Provided these stipulations are observed, we can very well have the Mahomet's tomb of legend suspended in mid air.

Finally, we can achieve this even through a magnetic *attraction*, provided we apply it to a moving object. This, incidentally, is the basic principle of a remarkable project for an electromagnetic, *frictionless* railway (Fig 95), suggested by the Soviet physicist Prof. B.P. Weinberg. I think it so instructive, that I have made so bold as to dwell on it in greater detail.

Electromagnetic Transportation

On Prof. Weinberg's projected railway the carriages *weigh nothing*, their weight being offset completely by electromagnetic attraction. They neither roll along rails, float on water, nor hover in air. They have no visible means of support, being suspended by the invisible "cables" of powerful magnetic forces. They

Fig. 95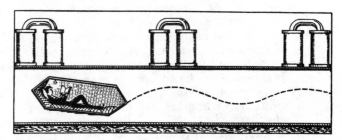

Frictionless railway designed by Prof. Weinberg.

experience no friction at all and, once set into motion, keep on moving by inertia. No locomotive is required for traction.

This is how the railway works. The carriages are placed in a copper tube that has been thoroughly evacuated to remove air resistance. As they *don't touch the sides of the tube*, they encounter no friction whatever. They are suspended in the middle of the airless tube by the attractive forces of powerful electromagnets, set at definite intervals on top, all along the tube outside, and capable of sending the *iron* carriages on their way through the tube without touching either "ceiling" or "floor". The electromagnet attracts the carriage moving beneath. But before it has a chance to bump into the "ceiling", it is pulled down by gravity, to be lifted again by the next electromagnet, before it reaches the "floor", thus racing along a smooth wavy trajectory, without friction and jolts. The carriages go through the void much like a planet in space.

The carriages are zeppeline-type cylinders of some 2.5 m long and about 90 cm tall. They are airtight, of course—since they move in a vacuum—and are outfitted with automatic submarine-type air-regeneration systems. They pull out in an entirely novel fashion that can compare only to a cannon shot. They are indeed "shot out" like cannon balls, the difference being that the "cannon" in this case is an electromagnetic one, based on the property that an electrically energised solenoid is able to pull in an iron plunger with great rapidity—which is the greater, the bigger the coil and the stronger the current. It is this force that shoots out the carriages and since, as we have noticed, there is no friction inside the tube, they continue to move by inertia with the same speed, until stopped by the solenoid at the point of destination.

Here are a few details provided by the author of the project:

"In the experiments I conducted in 1911-13 at the Physics Laboratory at the Tomsk Institute of Technology, I used a copper tube 32 cm in diameter with electromagnets on top and a 10-kg car on a support below. The car was a piece of iron pipe with front and back wheels and a nose-cone to stop it by ramming into a board reinforced by a bag of sand. It could

216

not go faster than about 6 km/h because of restricted space and the fact that the tube was a circle 6.5 m in diameter. I contest, though, that with 3-mile-long solenoids at the point of departure, a speed of 800-1,000 km/h could easily be worked up and maintained without expending any energy, since there is no friction on floor or ceiling to overcome.

"Though the railway, particularly the copper tube, would be rather expensive, no funds would be required to maintain the initial speed or keep a staff of engine-drivers, conductors, etc., and the costs per kilometre would not exceed several thousandths to several hundredths of a kopek; meanwhile a two-tube railway would be able to take daily 15,000 passengers or 10,000 tons in one direction.

Martians Fight Earthmen

The Roman savant Pliny relates a story current in his day about a magnetic promontory somewhere in India which attracted all iron objects with unusual force. Any unfortunate mariner, whose ship passed close by, was doomed, as his boat would lose all its iron nails and fall to pieces. Later this became one of the Arabian Night legends.

This, of course, is a legend and nothing more. There are magnetic mountains, or rather mountains rich in magnetite. Recall the famous magnetic mountains in the Urals, where the blast furnaces of Magnitogorsk now stand. However, the power of attraction of such mountains is exceedingly faint. There never was on the face of the earth anything of the sort that Pliny describes. And if today we do build ships with no iron or steel parts, this is done not for fear of magnetic rocks but to facilitate the study of terrestrial magnetism.

The science-fiction novelist Kurt Lasswitz took Pliny's legend as the basis for a story about a master weapon which Martian invaders in the novel *The Two Planets* employ to fight the Earthmen. With this magnetic or rather electromagnetic weapon, the Martians were able to disarm the Earthmen's troops without even engaging with them. Here is the extract describing the "battle".

"The glittering array of horsemen galloped forward. It seemed as if their selfless resolve had finally compelled the formidable foe [the Martians—*Ya. P.*] to retreat. There was a new commotion among the enemy airships. They ascended into the sky as if about to yield.

"At the same time, however, a dark spreading pall that had just appeared above the field, descended. It unfolded like a table-cloth, hemmed in on every side by airships. As soon as the first row of cavalry came beneath the odd machine it reacted with diabolical rapidity. The air was rent with deafening cries of horror. The horses and their riders fell. Meanwhile the air filled with clusters of clattering and rattling pikes, swords and carabines, that whirled up and clung to the strange machine.

"The machine then veered away slightly and dropped its crop of iron onto the ground. It returned twice seeming to collect every weapon in the field. None of the cavalrymen could hold his sword or pike.

"The machine was a new Martian invention. It attracted with an irrepressible force every steel and iron object. The Martians employed their hovering magnet to wrest from the enemy all their weapons, without doing them any harm.

"The aerial magnet floated towards the rows of infantry. In vain did the soldiers clutch at their rifles. Some invincible power plucked them out of their hands. Many who did not let go were themselves lifted into the air. The first regiment was completely disarmed in the matter of a few minutes. The machine then raced after the regiments marching into the city, ready to mete out the same treatment.

"Nor was the artillery spared."

Watches and Magnetism

Haven't you the impulse to ask: Couldn't a screen impenetrable to magnetic attraction be found? It could, of course. The Martians' fantastic invention could have been countered had due precautions been taken.

Queerly enough the substance impermeable to magnetic forces is that very same so easily magnetised iron. *Inside* an iron ring a compass needle will not be deflected by a magnet outside the ring. An *iron case* will protect the steel mechanism of a watch from magnetism. If you put a gold watch on the poles of a powerful horse-shoe magnet, all its steel parts, and, firstly, its fine hair-spring, will be magnetised' (provided, of course, that the hair-spring is not made of a special alloy called *invar* which does not magnetise, even though it contains both iron and nickel), and your watch would no longer show the right time. You wouldn't be able to repair the harm done, even if you took the magnet away; the steel mechanism would stay magnetised and your watch would need to be overhauled fundamentally. So don't experiment with a gold watch in this way. You'll have to pay dearly for it.

However, you can boldly perform this experiment with a watch whose mechanism is contained in an iron or steel case, as these two metals are impervious to magnetic forces. Even if you put it near the coil of

Fig. 96

What protects the steel mechanism of a watch from being magnetised?

a powerful dynamo, it will tick on just as well as it did before. These cheap iron-cased watches are ideal for the electrical engineer or technician.

Magnetic "Perpetual Motion" Machine

In attempts to invent a "perpetual motion" machine the magnet and its powers have played a role of no mean magnitude. Ill-starred "perpetual motion" machine inventors have tried might and main to apply the magnet to this end. Here is one such project (described back in the 17th century by the Englishman John Wilkins, the Bishop of Chester).

A powerful magnet A is placed on top of a pillar (Fig. 97), leaned against which are two inclined grooves M and N, one above the other. The upper groove M has a small hole C at the top, while the lower groove N is curved. The inventor presumed that the arrangement would operate as follows. A small iron ball B was to be placed on the upper groove. Attracted by magnet A, it ought to roll upwards. On reaching the hole, it should fall through on to the lower groove N, along which it should roll down, be carried up by inertia along the curve D, and find itself again on the upper groove M, from whence, again attracted by the magnet, it should again roll up and drop through the hole, roll down and on to the upper groove, *ad infinitum*. This, the inventor conjectured, would produce "perpetual motion".

Fig. 97

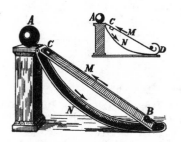

A fake "perpetual motion" machine.

Where did the inventor go wrong? His fallacy is easy enough to indicate. Why did he think that after rolling down groove *N* the little ball would have enough momentum to climb up curve *D*? This would happen, were the little ball influenced by gravity alone. However, we also have a second force, that of magnetic attraction, which is so powerful that it even compels the ball to ascend from position *B* to position *C*. Consequently, our little ball will not accelerate as it goes down groove *N*. On the contrary, it will roll very slowly down and, even if it should reach the bottom, will not have enough momentum to take the curve.

Sundry modifications of this same project have been put forward time and again. Curiously enough, one even had a patent issued to it. It happened in Germany in 1878, thirty years after the law of the conservation of energy was published. The inventor had so deftly guised the absurd principle at the root of his magnetic "perpetual motion" machine that he fooled the patent authority, though according to the regulations no patents are issued for inventions based on principles contrary to the laws of nature. However, the proud possessor of this sole patent of its kind ever issued, must have soon grown disillusioned, as he stopped paying patent duties in two years' time. Now anyone can acquire the "invention". However, I don't think anybody would ever need it.

Museum Problem

Museum experts often have to decipher ancient scrolls. They are so fragile that they tear though the utmost care is taken to separate the pages. The problem was how to perform this with success.

The USSR Academy of Sciences has a special document-restoration laboratory which tackles problems of this nature. In this particular case electricity was invoked. An electric charge was imparted to the MSs.

The pages received a unipolar charge and neatly separated without tearing—as like charges repel one another. After this it is simple for a proficient pair of hands to take the pages apart and mount them.

One More Fake "Perpetual Motion" Machine

The idea of coupling a dynamo to an electric motor has become very popular of late with all who want to "solve" the problem of perpetual motion. Every year I'm asked to advise on around half a dozen of these projects, which all boil down in essence to running a belt from the electric motor to the dynamo and wiring the dynamo with the motor. The idea is that if the dynamo be given an initial impulse, the electricity it generates will work the motor, which in turn will set the dynamo into motion. Consequently, as the inventors presume, the two machines will set each other in motion and never stop, until they both wear out.

It seems very tempting, doesn't it? But all who have tried it have found, to their surprise, that it doesn't work—as really was to be expected. Even if both machines had a 100-per cent efficiency factor, they would operate endlessly only in the complete absence of friction. This combination of machines— or assembly, as an engineer might call it—is really one machine which is supposed to make itself go. Of course, in the absence of friction it would move endlessly, just like anything else, but no practical use would ever be derived from it, because as soon as you tried to make it perform some work, it would stop at once.

This would give us "perpetual motion" but not a "perpetual motion machine". All that has just been said would be valid, naturally, were there no friction; since we do have friction, the machine won't go at all.

I wonder why these "perpetual motion" cranks never think of simply joining two pulleys by a belt and spinning one of them. Because, judging by the method of reasoning employed to "justify" the above-mentioned combination, oughtn't we to expect the first pulley to turn the second pulley, and the second pulley the first one? Why, we could even dispense with the second one. Wouldn't it be enough to turn just one pulley for its right half to cause the left half to turn and this left half similarly to cause the right half to turn?

This is so obviously absurd that I doubt whether they would ever bring a glint to the eyes of any "perpetual motion machine" inventor, though he labours under the same delusion.

A "Near-Perpetual Motion" Machine

The mathematician will, I think, pooh-pooh the idea of "near-perpetual motion". Either it is perpetual motion or not. "Near-perpetual motion" is actually not perpetual motion. But from the practical angle this can be viewed differently. I believe many would be quite satisfied to have a "near-perpetual motion" machine able to run for at least a thousand years. Man's life span is short and we would look upon a thousand years as eternity. I suppose people with a practical turn of mind would consider the "perpetual motion" machine problem solved, could that be done.

It can be; a thousand-year motion machine has been invented. Everybody can own one, provided he is willing to foot the bill. No patent has been taken for it and it is no secret. Commonly called a "radium clock", it was invented by Prof. Stret in 1903. It is rather simple (Fig. 98). It consists of a small glass tube *A*, with several thousandths of a gramme of *radium salts*, suspended from a quartz thread *B* (quartz does not conduct electricity), inside a sealed glass jar, from which the air has been evacuated. Attached to one end of the tube, as in electroscope, are two narrow strips of gold leaf. Radium, as you may know, emits three types of rays—respectively, alpha, beta and gamma rays. In our case it is the beta rays, as easily capable of passing through glass and comprising a stream of negatively charged particles or electrons, that do the job. The scattered electrons carry away the *negative* charge gradually imparting a *positive* charge to the radium in the tube. This positive charge is then imparted to the strips of gold leaf, causing them, on account of mutual repulsion, to swing apart, touch the jar's wall, relinquish their charge to strips of electricity-conducting tin foil, appropriately past-

Fig. 98

A radium clock with an almost "perpetual wind" for 1,600 years.

ed on, and swing together again. With every new electron charge, this process repeats itself—which happens every two or three minutes with the regularity of a pendulum. Hence the name. This continues for years and centuries until the radium decays entirely. This of course is not a "perpetual motion" machine but merely a "gift-motion" machine.

How long does radium emit rays? Scientists have found that radium has a half-life of 1,600 years. Consequently, a radium clock will be able to go without stopping for at least a thousand years. Only the frequency of oscillations will diminish due to the diminishing electric charge. Had a radium clock of this order been set going when Russia was born as an independent state, it would still be ticking today!

Could this "gift-motion" machine be used for any practical purposes? Unfortunately, not. Its power, or the amount of work it does every second, is so minute that it will never set any mechanism into

225

motion. To attain tangible results, we would need a far larger stock of radium. Since radium is an extremely rare and very costly element, such a "gift-motion" machine would be ruinous.

Insatiable Birdie

There is a Chinese toy which is a perpetual source of astonishment and delight. This is the "insatiable birdie". Put before a drinking bowl, the "birdie" will dip its beak in the water and having "drunk its fill", swing back into its initial upright position. After a

Fig. 99

Insatiable birdie.

while, it slowly leans over to dip its beak in the water again, "drinks" and swings back. This is a typical "gift-motion" machine, one, moreover, based on an extremely ingenious principle. Look at Fig. 99. The birdie's "body" consists of a glass tube ending at the top in a little head-shaped sphere, and inserted at the bottom into a broader similarly air-tight reservoir filled with ether up to a level a bit higher than the open bottom end of the tube.

To cause the birdie to "drink", its head must be wettened with water. For a while we see it still upright—since the bottom reservoir with the ether is heavier than the head. Then we see the ether gradually rise in the tube (Fig. 100), finally making the upper part heavier and causing the birdie to lean over and dip its beak in the drinking bowl. When the birdie swings into a horizontal position the tube's open

Fig. 100

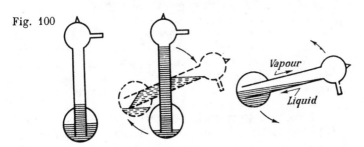

The birdie's secret.

bottom end takes up a position which is above the level of the ether in the bottom reservoir, thus causing the ether in the tube to flow back into the tail reservoir, and the birdie, consequently, to swing back into its initial upright position. This is the mechanical aspect of the problem; as the ether rises and returns, the centre of gravity shifts.

But what compels the ether to rise? Ether easily evaporates at room temperature, while the pressure that saturated ether vapours exert greatly varies, when the temperature fluctuates.

227

When the birdie is in its upright position, there are two clearly distinguishable zones of ether vapours. There are the "head" tube and the "tail" reservoir.

Now the "head" possesses the remarkable property of lowering its temperature in comparison with that of the surrounding environment. This can be achieved by making the head of a porous material well able to soak in moisture and intensively evaporate it.

In Chapter Seven I told you that intensive evaporation causes a drop in temperature. Since the head-tube's temperature becomes lower than that of the tail reservoir, this causes a drop in the pressure of the saturated vapours in the head-tube, thus compelling the ether to be forced up the tube by the greater pressure of the vapours in the tail reservoir. The centre of gravity shifts and the birdie swings into a horizontal position. Then two absolutely independent processes take place. In the first place, the birdie wettens its beak and thus soaks its cotton wool head. Secondly, the saturated vapours in the head-tube mix with the vapours in the tail reservoir. Pressure evens out (owing to the heat of the surrounding air, the temperature of the vapours rises slightly), and the ether flows back by gravity into the tail reservoir, causing the birdie to swing back into its upright position again.

This "nodding" continues while the cotton wool head is wet and provided the humidity of the surrounding air is not too great.

These two factors will guarantee normal evaporation and the consequential relative drop in the head-tube's temperature. It is thus that the warmth of the surrounding air imparted to the birdie all the time causes it to nod.

228

A study of the laws governing the decay of radioactive elements has provided scientists ·with a reliable method for estimating the earth's age.

What is radioactive decay? It is the "spontaneous transformation" of one set of atoms into another, a process not induced by any external factors. It is curious to note that temperature and pressure fluctuations and the like have not the slightest effect on the "speed" of this process (only a temperature of the order of tens of thousands of millions of degrees could have any effect). The elements of uranium, thorium, and actinium, contained in some minerals, are the progenitors of several series of radioactive elements, each of which series is a sequence of spontaneous transformations of radioactive elements. The ultimate product of these spontaneous transformations in the case of all three elements named is lead. True, this "lead" in each case has a different "atomic weight" than the normal lead. A normal atom of lead is 207 odd times heavier than an atom of hydrogen, while atoms of lead closing the uranium, thorium and actinium series are respectively 206, 208 and 207 times heavier. It is thus quite possible to distinguish between them.

This spontaneous transformation is attended by the emission of what are called alpha rays—which are a beam of charged particles of matter or atoms of the light inert gas helium—by the decaying atoms. As these charged particles possess enormous velocities, when released they lose their positive charge and settle down in the mineral as particles of ordinary helium— which explains why we find helium in all radioactive minerals.

229   However, the estimation of a mineral's age by

its helium content may lead us very far out, since helium—like any light gas—evaporates.

Couldn't a more accurate estimate of age be obtained by checking the amount of lead accumulated in the mineral? In the early 1940s, proceeding from a quantitative evaluation of lead isotopes at different deposits, the British geologist Holmes concluded that the earth was 3,500 million years old. Actually he had ascertained the age not of the earth but of its crust and had furthermore proceeded from the obsolete conjecture that the earth owed its origin to a hot clot of gas torn out of the sun.

In 1951-52 Academician A. P. Vinogradov concluded, after scrupulously analysing all available data, that it was impossible to estimate the age of the earth's crust on the basis of lead data alone. One could only surmise that it was not more than 5,000 million years old. At the same time minerals were found, whose age has been estimated as 3,000 million years. Proceeding from data concerning the speed with which identical quantities of two uranium isotopes (with the atomic weights of 235 and 238) decay, scientists have put the earth's age between 5,000 and 7,000 million years.

So we may presume that the earth is some 6,000 million years old. This estimate seems to be right because the same figure was deduced, though entirely different methods were used. Naturally 6,000 million years is tremendous compared even with man's history, let alone his life-span.

Birds on the Wires

You all know how dangerous it is for a person to touch live tramway or high-tension wires. We know of many cases too, when people and big animals have been electrocuted by a broken live wire. Why then do birds so calmly perch on live wires with absolute impunity (Fig. 101)?

To understand this contradiction the following must be taken into account. One must regard the body of

Fig. 101

Birds perch with impunity on electric wires. Why?

the bird, perched on the wire, as the branch line of an electric circuit, whose resistance is tremendous compared with the other branch line of the circuit, namely, the very short distance between the bird's claws.

Therefore the intensity of the current passing through the bird's body is negligible and, consequently, harmless. But as soon as the bird on the wire touches the pole with its wing, tail, or bill, or makes a contact with the ground in any other way, it is immediately electrocuted. No doubt you have observed this.

Birds have the habit of pecking at the live wire while perched on the supports of high-tension transmission lines. Since the support is not insulated from the ground, the moment the grounded bird touches the live wire it is electrocuted.

This happens so often that the Germans once took special bird-protection measures installing insulated perches on the supports of high-tension transmission lines, which thus prevented electrocution when the bird chose to peck at the live wire (Fig. 102). Some-

times, to keep the birds away, live wires are enclosed
in guards.

Fig. 102

An isolated bird perch on
the support of a high-vol-
tage transmission line.

In the Light of Lightning

Have you ever happened to observe busy street
crossings when a thunderstorm flashes? Imagine for a
moment that a thunderstorm has caught you out shop-
ping. Oddly enough, during the flashes of lightning
everything, so full of life a moment ago, seems motionless; cars seem to be standing still and you are able
to clearly make out every spoke in their wheels.

The reason lies in the negligible fraction of time
during which lightning flashes. Like any electric
spark lightning is instantaneous, so much so that our
usual methods of time-reckoning simply won't do.
However, scientists have indirectly established that
a flash of lightning lasts less than a thousandth of a
second.* In such extremely short stretches of time

---

* There are also cases of flashes of lightning lasting for "as
long" as several tens of a second, and of a series of repeated
flashes, each bolt whizzing through the "corridor" made by
the first, that will keep up for "as long" as one and a half seconds.—*Ed.*

scarcely anything moves to any noticeable degree. No wonder that in a flash of lightning a busy street seems motionless. After all we see only what lasts less than a thousandth of a second. In this time interval the wheel of a car shifts only a negligible fraction of a millimetre—which our eye cannot catch at all.

How Much Does Lightning Cost?

In those ancient times when lightning was the prerogative of "divine beings" a question of this order would have been regarded as rank blasphemy. Today, however, when electricity is a commodity that can be gauged and valued like any other, I don't think anyone would take the question of the cost of lightning as bosh and bunkum. Our problem is to reckon the amount of electric energy consumed in a lightning discharge and price it at least on the basis of electricity rates.

The potential of a lightning discharge is about 50 million volts, the maximum intensity of current rating 200,000 amperes (a factor ascertained incidentally by noting the degree to which a steel rod is magnetised by the current that passes through its coil when lightning strikes at a lightning conductor). The wattage is determined by multiplying the number of volts by the number of amperes. As we do this, however, we must not lose sight of the fact that during the discharge the potential drops to zero. Consequently we must take the mean potential or, in other words, half the initial voltage. This gives us discharge wattage of $\frac{50,000,000 \times 200,000}{2} = 5,000,000,000,000$ watts, or 5,000 million kilowatts.

This impressive row of naughts may lead us to believe that we shall have a similarly astronomical figure representing cost. However, to get kilowatt-hours—which electricity rates are based on—the time factor must also be taken into account. Our lightning discharge lasts about a thousandth of a second, in which interval $\frac{5,000,000,000}{3,600 \times 1,000}$ or about 1,400 kwh are spent. The rate of a kilowatt-hour is 4 kopeks and hence lightning should cost $1,400 \times 4 = 5,600$ kopeks or 56 rubles.

It's astounding, isn't it! Lightning, the energy of which is a hundred times greater than that released when a heavy artillery piece is fired costs but 56 rubles.

It is interesting to note how close electrical engineering has got to the artificial manufacture of lightning. In laboratories scientists have produced a tension of up to ten million volts and have obtained a fifteen-metre-long spark.

Thunderstorm at Home

It is very easy to make at home a small fountain by using a rubber tube with one end either in a pail of water that has been placed on a chair, or attached to a tap. The other end of the tube must be very narrow if you want to get the fountain to spray. The simplest way to do that is to make a nipple by using a pencil stub from which the lead has been extracted. For convenience's sake you may insert the pencil stub in the end of an overturned funnel as shown in Fig. 103.

Get your fountain to play to about half a metre high, with the spray directed towards the ceiling. Then bring near a stick of sealing wax or an ebonite comb previously rubbed with a piece of cloth. The separate fountain sprays will suddenly merge into a solid column of water that thuds quite loudly on a plate with a sound exceedingly reminiscent of the noise of a thunderstorm. "Unquestionably," the physicist Boyce remarks in this connection, "this is precisely why the raindrops in a thunderstorm are so large". As soon as you withdraw the sealing wax or

Fig. 103

Thunderstorm in miniature.

comb, your fountain will spray again and you will hear the gentle tapping of scattered drops of water.

You may turn this into a parlour trick and cause the uninitiated to believe that your stick of sealing wax is a "magic" wand.

The phenomenon is caused by the induced electrification of the drops of water. The drops near the stick receive a positive charge and the drops further away, a negative charge; mutual attraction thus causes the drops to merge.

There is a much simpler way of detecting the effect electricity has on a jet of water. Put a comb through your hair and then bring it up to the trickling from a tap. The trickle will become a steady stream and curve noticeably (Fig. 104). This is harder to explain than the previous phenomenon; it is bound up with the change in surface tension under the effects of an electrical charge.

Let me note in passing that the ease with which an electric charge accumulates during friction explains the electrification of belts as they rub against pul-

Fig. 104

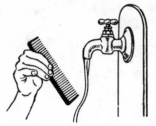

The stream of water from the tap bends away when an electrified comb is brought up.

leys. The electric sparks thus emitted sometimes cause fires. To avoid this the belts are thinly plated with silver. This turns them into conductors of electricity, unable to accumulate an electric charge.

Chapter 9 *Reflection and Refraction of Light. Vision*

Quintuple Photo

Fig.105

A quintuple photograph of one and the same face.

Fig. 105 shows you a photographic curio—it depicts one and the same person in five different ways. Photos of this sort are unquestionably much better than ordinary snapshots since they provide a much fuller picture of a person. You all know the great pains a photographer will take to get the most characteristic turn of the head. A "quintuple photo" gives you several from which you may choose the best.

How are these photographs taken? With the help of mirrors (Fig. 106). The sitter, who has his back

Fig. 106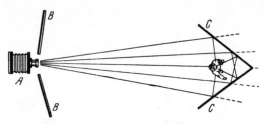

How to obtain multiple photographs. The sitter is placed between the mirrors CC.

turned to camera A, faces two upright flat mirrors CC, set at an angle of one-fifth of 360°, that is, 72°. This pair of mirrors should give four reflections turned differently towards the camera. It is these reflections, plus the sitter, that are photographed. The unframed mirrors, naturally, do not show. To prevent reflection of the camera itself it is screened off by two screens BB which are placed so as to leave a small aperture for the lens.

The number of reflections depends on the angle between the two mirrors; the acuter the angle, the greater the number of reflections. If the angle is 90°, which is 360:4, we get four images; if it is 60°, which is 360:6, we get six. Should the angle be 45°, which is 360:8, we get already eight images, and so on and

238

so forth. However, the greater the number of reflections the fainter the image is. That is why photographers usually prefer to confine themselves to a fivefold reflection.

Sun-Powered Motors and Heaters

The idea of employing solar energy to heat an engine boiler is most fascinating. A simple reckoning: scientists know how much energy a minute every square centimetre of the top surface of the atmosphere, in places at right angles to the sun's rays, receives from the sun. Since the amount is apparently constant, it has come to be known as the "solar constant". In round numbers it is equal to two calories per square centimetre a minute. This amount of heat which the sun sends us with such faithful regularity does not all reach the earth's surface. About half a calorie is absorbed in the atmosphere. We may take it for granted that every square centimetre of ground placed perpendicular to the sun's rays gets roughly 1.4 calories a minute. Converted into square metres, this means 14,000 calories or 14 large calories a minute, or about $1/4$ of a large calorie a second. Since one large calorie gives 427 kgm of mechanical work, sunshine falling perpendicularly on one square metre of ground could provide more than 100 kgm of energy per second, or in other words more than $1\frac{1}{3}$ power.

Such is the amount of work solar energy could do in optimal conditions, i.e., when its incidence is perpendicular and it is all converted into mechanical energy. However, we have a long way to go yet to achieve this ideal. The efficiency factor obtained so far has not been more than 5-6 per cent, as a rule. The best efficiency factor of 15 per cent has been recorded by Prof. Charles Abbot's sun-powered motor.

It is much easier to employ solar energy not for mechanical work but for heating. In the USSR great attention is paid to this problem. There is a special Sun Institute in Samarkand, which carries on extensive research. Various solar devices, including bathhouses and water heaters, have been devised and

Fig. 107

Sun-powered water-heating unit in Turkmenia.

tested. The average efficiency factor of solar water heaters is 47 per cent, while the maximum is 61 per cent. One such device tested in Turkmenia was a solar refrigerator, whose cooling batteries ran a temperature of 2-3°C below zero, despite an air temperature of 42°C above zero in the shade. This is the first commercial solar refrigerator of its kind.

Experiments in the sun-powered smelting of sulphur—which has a melting point of 120°C—produced some very fine results. I could also mention solar distillers used along the Caspian and Aral seas to obtain fresh water, a sun-powered pump to replace the primitive Central Asian pumps, sun-driers for drying fruit and fish, a solar kitchen-range, etc. This is a far from complete list of all the different ways

Fig. 108

Sun-powered cold storage in Turkmenia.

in which trapped sunshine may be used. It will definitely play an important role in the economy of Central Asia, the Caucasus, the Crimea, the Volga delta and the Southern Ukraine.

Cap of Invisibility

There is a hoary legend of a magic cap which renders the wearer invisible. In his *Ruslan and Ludmila*, the celebrated Russian poet Alexander Pushkin provides a classical description of the miraculous properties of this cap:

> *It came into the maiden's mind—*
> *As happens oft with womankind—*

241

To try the wizard's head-piece on.
Ludmila twists it to and fro;
Straight, sideways did she try to don
The wizard's wondrous cap, and lo!
When she had tried it back-end foremost—
Amazement would have killed you almost—
Ludmila vanished out of sight!
But then again she put it right
And reappeared. On repetition
Ludmila disappeared anew.
"A-ha, my wily old magician,
I owe much gratitude to you.
For hence my freedom I have found
And shall be ever safe and sound!"

The cap of invisibility was all the captive Ludmila had for protection. She used it to escape the clutches of her Argus-eyed guards, who were able to suspect her presence only by her action.

Many of the wonders of ancient legend have long become facts of everyday life. Science has wrought many a miracle. Today we can drill mountains, trap lighting, and travel by "flying carpets". Couldn't we invent also the cap of invisibility or find some means of making ourselves absolutely invisible? Let us discuss this point.

The Invisible Man

In his *The Invisible Man* H. G. Wells would have us believe that it is quite possible to render ourselves invisible. The main character in the novel, whom the author calls the most brilliant physicist the world had ever known, discovered a way of making the human body invisible. In the following passage he describes his method to a doctor acquaintance.

"... visibility depends on the action of the visible bodies on light... You know quite well that either a body absorbs light or it reflects or refracts it or does all these things. If it neither reflects or refracts nor absorbs light, it cannot of itself be visible. You see an opaque red box, for instance, because the colour absorbs some of the light and reflects the rest, all the red part of the light to you. If it did not absorb any particular part of the light, but reflected it all, then it would be a shining white box. Silver! A diamond box would neither absorb much of the light nor reflect much from the general surface, but just here and there where the surfaces are favourable the light would be reflected and refracted, so that you would get a brilliant appearance of flashing reflections and translucencies. A sort of skeleton of light. A glass box would not be so brilliant, not so clearly visible as a diamond box, because there would be less refraction and reflection ... if you put a sheet of common white glass in water, still more if you put it in some denser liquid than water, it would vanish almost altogether, because light passing from water to glass is only slightly refracted or reflected, or indeed affected in any way. It is almost as invisible as a jet of coal gas or hydrogen is in air. And for precisely the same reason!

"'Yes,' said Kemp. 'That is plain sailing. Any schoolboy nowadays knows all that.'

"'And here is another fact any schoolboy will know. If a sheet of glass is smashed, Kemp, and beaten into a powder, it becomes much more visible while it is in the air; it becomes at last an opaque, white powder. This is because the powdering multiplies the surfaces of the glass at which refraction and reflection occur. In the sheet of glass there are only two surfaces, in the powder the light is reflected or refracted by each grain it passes through, and very little gets right thro-

Fig. 109

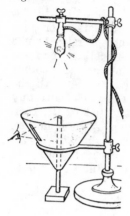

ugh the powder. But if the white, powdered glass is put into water it forthwith vanishes. The powdered glass and water have much the same refractive index that is, the light undergoes very little refraction or reflection in passing from one to the other'.

"'You make the glass invisible by putting it into a liquid of nearly the same refractive index, a transparent thing becomes invisible if it is put in any medium of almost the same refractive index. And if you will consider only a second, you will see also that the powder of glass might be made to vanish in air, if its refractive index could be made the same as that of air. For then there would be no refraction or reflection as the light passed from glass to air.'*

---

* We can render an absolutely transparent object completely invisible if we surround it on all sides by walls able to disperse light strictly uniformly. Through a small side aperture our eye will then receive from every point of the object the same amount of light and the object will seem nonexistent, as it will have no sparkle or shadow betraying its presence. Here is how you do it. Make a funnel out of white carboard half a metre in diameter and mount it as shown in Fig. 109 some

"'Yes, yes,' said Kemp. 'But a man's not powdered glass.'

"'No,' said Griffin. 'He's more transparent.'

"'Nonsense.'

"'That's from a doctor. How one forgets! Have you already forgotten your physics in ten years? Just think of all the things that are transparent and seem not to be so! Paper, for instance, is made up of transparent fibres, and it is white and opaque only for the same reason that a powder of glass is white and opaque. Oil white paper, fill up the interstices between the particles with oil, so that there is no longer refraction or reflection except at the surfaces, and it becomes as transparent as glass. And not only paper, but cotton fibre, linen fibre, wool fibre, woody fibre, and *bone*, Kemp, *flesh*, Kemp, *hair*, Kemp, *nails* and *nerves*, Kemp, in fact, the whole fabric of a man, except the red of his blood and the dark pigment of hair, are all made up of transparent, colourless tissue—so little suffices to make us visible one to the other...."

Confirmation is afforded by the considerable transparency of the tissues of nonhairy albino animals (their tissues are completely devoid of pigment). One zoologist who happened to pick up an albino frog near

---

distance away from a 25-W bulb. Then insert from below through the funnel's narrow end a glass rod, keeping it strictly vertical, as the slightest deviation from the vertical will cause it to appear either as a haloed dark thin shadow or as a shadow-fringed pencil of light, interchanging as soon as you give the stick a slight twist. By trial and error get the light to fall uniformly on the stick. Then to the eye, looking at it through a narrow side aperture not more than 1 cm in diameter, it will *seem nonexistent*. The glass object will be completely invisible, though its refractivity greatly differs from that of air. Another way of making a bit of faceted diamond-cut glass invisible, is to place it in a box coated inside with luminescent paint.

Detskoye Selo in the summer of 1934, describes it as follows: "The thin skin and muscular tissues are transparent and one can see the skeleton and visceral organs through them. The contraction of the heart muscle and intestine movement are particularly well seen through the abdomen."

Wells' hero discovered how to make all the tissues of the human organism and even its pigments transparent. He applied his invention to his own body obtaining the spectacular result of total invisibility. Let us now see what happened next to the invisible man.

Might of Invisibility

Wells demonstrates with extraordinary wit and logic that an invisible man thus acquires almost unlimited power. He is able to enter any place unnoticeably and steal anything with impunity. Elusive, thanks to invisibility, he successfully fights a whole crowd of armed people. Threatening to smite all who are visible, the invisible man subjugated the population of an entire town. Himself elusive and invulnerable, he strikes down all his opponents despite their every precaution. The invisible man is thus able to issue to the terrified population of his home town an order of the following content:

"Port Burdock is no longer under the Queen, tell your Colonel of Police, and the rest of them; it is under me... This is day one of year one of the new epoch—the Epoch of the Invisible Man. I am Invisible Man the First. To begin with, the rule will be easy. The first day there will be one execution for the sake of example—a man named Kemp. Death starts for him today. He may lock himself away, hide himself away, get guards about him, put on armour if he likes—Death, the unseen Death, is coming. Let him take precautions—it will impress my people... Death starts. Help him not, my people, lest Death fall upon you also."

At the outset, the invisible man triumphs. Only with the greatest of difficulty do the terrorised townsfolk rid themselves of their invisible foe who dreamed of becoming their all-powerful master.

Transparent Preparations

Are the physical theses on which this science-fiction novel is based right? Unquestionably yes. In a transparent medium every transparent object becomes invisible when the difference between refractive indices is less than 0.05. Ten years after H. G. Wells' *The Invisible Man* was published, the German anatomist, Prof. W. Spalteholtz, put the writer's idea into practice—true not on living organisms but in the preparation of dead specimens. Such transparent preparations of organs and even whole animals may be seen today in many museums. The transparent preparations method evolved by Prof. Spalteholtz in 1911 is briefly as follows. After treatment—bleaching and washing—the prepared specimen is soaked in methylsalicylate, a colourless liquid with a big refractive index. Specimens of rats and fishes, or various human organs thus prepared, are placed in jars containing the same liquid. However, full transparency is not sought as this would cause the specimens to become absolutely invisible and, consequently, useless for the anatomist. We could achieve full transparency if necessary though.

This is naturally far from Wells' dream of a *live man* so transparent as to be absolutely invisible. Firstly, because we must know how to treat *living* tissue with this transparency liquid without violating organic functions. Secondly, because Prof. Spalteholtz's preparations are transparent but not invisible. They are invisible only while immersed in liquid of corresponding refractivity. They will become invisible in the air, only when their refractive index is the same as that of the *air*—which is something we are still unable to achieve.

However let us imagine for a moment that with time we shall be able to do this and consequently realise the British novelist's dream. H. G. Wells was so thorough, that one can't help believing him and his thesis that an invisible man must indeed be the most powerful of mortals. This is not at all so. There was one point which *The Invisible Man*'s clever author overlooked.

Can an Invisible Man See?

Had Wells ever stopped to ask himself this question before he embarked on his novel we would have never had the pleasure of reading his gripping narrative. This upsets the entire apple-cart because *an invisible man must be ... blind.*

Why couldn't the invisible man be seen? Because every part of his body, including his eyes, was rendered transparent and possessed a refractive index identical to that of the air.

Let us now recall the eye's function. Its crystalline lens, vitreous humour and other elements refract light so as to produce a retinal image of surrounding objects. But when the refractivity of the eye and that of the air are equal, the sole cause of refraction is removed. Passing from one medium to another of *the same refractivity*, light will not change its direction and, consequently, its rays will be unable to concentrate in one point. Light will pass through the eye of an invisible man without hindrance; its rays will neither be refracted nor retarded—since there is no pigment. To induce a sensation in animals, the rays of light must bring about some changes even of the most insignificant nature; or, in other words, perform certain functions in the eye. Consequently, at least part of the rays must be *retarded*. An absolutely transparent eye will naturally be unable to check rays, otherwise it wouldn't be transparent. All creatures drawing upon transparency for protection have eyes that are not completely transparent, provided they have eyes. "Immediately below the surface," the well-known oceanographer Murray writes, "most animals are transparent and colourless.... When taken from the tow-nets they are often distinguishable *only by their little black eyes*, their blood being devoid of haemoglobin and the entire body perfectly transparent—and consequently will fail to produce any mental image."

To sum up: *an invisible man sees nothing*. He will derive no benefit from all his advantages. This formidable claimant to power would have to grope in the darkness begging for alms which nobody would be able to give, as the supplicant would be invisible. Instead of the most powerful of mortals we would have before us a helpless cripple doomed to a miserable existence. (It is quite likely that Wells deliberately omitted this point. In his science-fiction, he often deliberately obscures the basic defect by plentiful realistic detail. In the preface to an American edition of his novels, he directly says that as soon as the trick has been done everything else must be made to seem common and lifelike. One must stake, he wrote, not on the power of logic, but on the created illusion.)

In other words, if we desired the cap of invisibility, it would be futile for us to copy Wells. Even a successful result would be a sorry one.

Protective Paint

There is, however, another way of acquiring the cap of invisibility. This is to paint objects so as to render them unnoticeable to the eye. Nature supplies countless instances, it draws extensively on this simple means to protect creatures from enemies and help them in their difficult struggle for existence.

What the military call camouflage, zoologists have been calling mimicry ever since Darwin. Fauna affords thousands of instances, we meet them at every step. Most desert denizens have the characteristic yellow colouring of sand—the lion, for instance, or any bird, lizard, spider or worm. Arctic inhabitants, on the contrary—be it the dangerous polar bear or innocuous bear or innocuous loon—all possess a natural white colour making them inconspicuous against the background of snow. Butterflies and caterpillars living on trees have a colouring that simulates bark with surprising accuracy—the praying mantle, for instance.

Every bug and insect collector knows how hard it is to spot specimens because of natural mimicry. Try to find a green grasshopper chirping away at your feet. You will never make it out against the green background.

The same is true of aquatic creatures. Marine animals living amidst brown seaweed all have a protective brown colouring. Among red seaweeds, the dominant protective colouring is red. The fish's silvery scales also serve to protect it both from birds of prey above and the carnivores of the ocean deep. The surface of water is mirror-like both from above and still more so from beneath—"total reflection"—and the fish's silvery scales well merge with this sparkling metallic background. Meanwhile jelly-fish and other transparent sea denizens such as worms, shellfish, molluscs, and the like, have chosen as protective colouring the total lack of colour, a transparency that renders them invisible in the colourless and transparent world which they inhabit.

Nature's devices are far superior to anything man has ever invented. Many animals are able to adapt their colouring to the changes of nature. The silvery-white ermine, so unnoticeable against the snow, would be easy prey were it not to change its colour when the snow melts. Every spring this white beast dons a new coat of a russety brown, to merge with the ground now bereft of snow, and goes white again in winter.

Camouflage

Nature's ingenuity has supplied a lesson or two in the art of camouflage, the use of blending or deceptive coloration. The dressy uniforms of the past, which made battle scenes so colourful, have been completely superseded by the familiar khaki. The steely grey paint of warships ("battleship grey") is also a protective colouring rendering ships inconspicuous on the seas.

In military camouflage we use branches, bizarre-painted designs, smoke and other deceptive expedients to disguise guns, forts, tanks, and ships. Camps are hidden under special nets with tufts of grass. Fighting-men put on masking capes.

Military aircraft are likewise camouflaged by having the top painted brown, a dark-green or violet—to blend with the ground and thus be rendered invisible to spotters from above—and the belly daubed a pale blue, pink, or white—to blend with the sky and thus become unnoticeable to ground observers. At 750 m aircraft thus camouflaged are hardly noticeable while at 3,000 m they become invisible. Night bombers are painted black.

An ideal protection suitable for *any* occasion is a *mirror-like* surface capable of reflecting the environment. Then the object automatically becomes mimetic and it is practically impossible to discern it from a distance. The Germans employed this method to camouflage their Zeppelins in the First World War. Their glistening aluminium body reflected the skies and the clouds, making them extremely difficult to spot—unless betrayed by engine noise.

So has the fairy-tale legend of invisibility been realised in practice in nature and warfare.

Underwater Eye

Assume, for a while, that you were able to stay under water for as long as you liked and could keep your eyes open there. Would you be able to see anything? You might think that since water is transparent nothing should prevent you from seeing just as well as you do in air.

However, recall the blindness of the "invisible man". He cannot see because the refractive indices of his eyes and the air are the same. Under water we find ourselves in roughly the same conditions as the "invisible man" in the air. A few figures to make things clear. The refractive index of water is 1.34. The refractive indices for the transparent media of the human eye are: 1.34 for the vitreous humour and cornea; 1.43 for the crystalline lens, and 1.34 for the aqueous humour. As you see the refractivity of the lens is only a tenth greater than that of water, while as far as the other elements of the eye are concerned the refractive indices are *identical*. That is why under water the rays focus far behind the retina, thus producing an extremely blurred retinal image. Only very short-sighted people can see more or less normally under water.

If you would like to get a clear picture of the way things would appear under water put on a pair of *biconcave* spectacles. These lenses which strongly dis-

Fig. 110

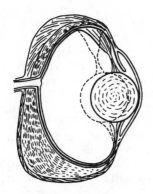

Cross-section of a fish's eye. The crystalline lens is spherical in shape and it does not change its form during accommodation. Instead, the position changes, as shown by the dotted line.

perse light will focus all the rays that the eye refracts way beyond the retina, thus producing an extremely blurred image.

But couldn't a person under water use glasses with a big refractive index? Ordinary lenses won't help much, as the refractive index of ordinary glass is 1.5—just a little more than that of water (1.34) and under water their refractibility will be feeble. One must have special glasses with very great refractivity (so-called heavy flint glass has a refractive index of almost 2). With such glasses you would be able to see more or less distinctly under water (read further about special goggles for divers).

253

Now you will have realised why a fish's crystalline lens is extremely concave, in fact, spherical; its refractive index is the greatest of all for animal eyes. Were this not so, it would be useless for fish as inhabitants of a greatly refracting transparent environment to have eyes at all.

How Do Divers See?

Many will probably ask: how are divers in their diving suits able to see anything under water, when our eyes hardly refract in the water at all? After all, diving helmets are always equipped with flat—not convex—glasses. Then how could the travellers in Jules Verne's *Nautilus* admire under water scenery through the portholes?

This is easy enough to answer. You must realise that when we dive in without a diving suit and helmet, the water comes into *direct* contact with our eye. In a diving helmet, however (or inside the *Nautilus*) the eye is *separated from the water by a layer of air* (and glass), which essentially changes matters. Emerging from the water and passing through the glass, light first goes through air before it reaches the eye. In conformity with the laws of optics the rays of light from the water that strike the *flat parallel glass* at an angle *do not change their direction* as they go through the glass. But as soon as they pass from the air into the eye, they naturally refract, with the eye performing the same functions as it performs normally. A fine illustration is that we experience no trouble at all in seeing fish in a fish bowl.

Lenses under Water

Have you ever tried to immerse a magnifying glass in water and look at submerged objects through it?

Surprisingly enough in water the magnifying glass hardly magnifies at all. If you were to do the same with a biconcave lens, you would again find that it had lost its powers of diminishing the size of objects. Furthermore, if you were to use not water but a liquid with a refractive index greater than that of glass, a biconvex lens would *diminish* the size of objects, while a biconcave one, on the other hand, would *magnify* them.

You need merely recall the law of the refraction of light to understand why. A magnifying glass magnifies because the lens refracts light to a greater degree than

Fig. 111

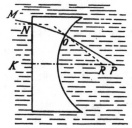

Divers' goggles are hollow lenses, one side of which is flat and the other concave. As it refracts, the ray *MN* follows the path *MNOP swinging away* inside the lens from the normal incidence and *veering back* to it (i.e., to *OR*) outside the lens. That is why the lens acts as a collecting glass.

the surrounding air. However, since the difference between the refractivity of lenses and water is very small, light does not bend much when crossing from the water into the lenses. This is the explanation for the powers of lenses, whether biconvex or biconcave. Monobromo-naphthalene, for instance, has a *greater* refractive index than glass and that is why convex lenses reduce the image in it, while concave ones magnify it. Hollow, or rather air-filled lenses, act similarly under water; concave lenses magnify while convex lenses reduce. Diving goggles are comprised precisely of such air-filled lenses (Fig. 111).

Such people often run great risks merely out of disregard for one noteworthy consequence of the law of refraction. They do not know that refraction seems to elevate everything in the water above its true position. To the eye the bottom of a pond or river seems *raised* by nearly a *third of the depth*. This delusion often places people in great peril. It is especially important for children and persons of small height to be aware of this, for otherwise a mistake in gauging the depth may prove fatal. The same optical principle of refraction that distorts the image of a spoon in a teaglass (Fig. 112), produces this apparent elevation of the bottom of a pond.

Refraction may be illustrated as follows. Sit your friend at the table so that he cannot see the bottom of a bowl placed in front of him. Place a coin on the bottom of the bowl so that it is hidden from his eye by the side of the bowl. Now ask him not to bend over, and pour some water in the bowl. To his surprise the coin will come completely into view. Remove the water with a syringe and the coin will disappear (Fig. 113).

Fig. 114 provides the explanation. To the observer's eye at point A above the water, coin m seems elevated. The rays are bent and, passing from the water to the air, enter the eye as shown in the figure. The eye thus imagines the coin to be where these lines continue, that is, above its true position. The more the rays are slanted, the higher the coin seems to elevate. That is why, when looking from a boat at the even bottom of a pond, we always think it deepest right beneath us and shallower further off, or, in other words, concave.

256   On the contrary, if we were able to look from the

Fig. 112

Distorted image of a spoon in a glass of water.

Fig. 113

Coin-in-bowl experiment.

Fig. 114

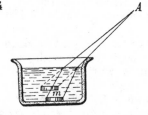

Why the coin in the experiment illustrated in Fig. 113 seems elevated.

Fig. 115

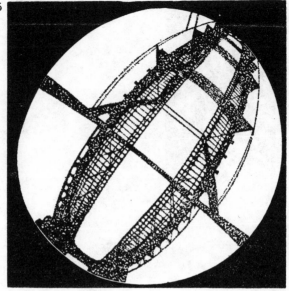

What a railway bridge across
a river would look like to
an underwater observer.
(From a photograph made by
Prof. Wood.)

bottom of a pond at the bridge spanning it, the bridge
would seem convex (as shown in Fig. 115; later I shall
tell you how this photograph was taken). In this
particular case the rays pass from a medium with a
poor refractivity (the air) into one with good refractivity (the water); that is why the effect is opposite
to what we see when the rays pass from water into
air. For the same reason a fish, looking at a row of
people outside a fishbowl, should see them as standing
not in a straight line but in an arc with the bulge

facing it. A bit later I shall tell you in greater detail
how fish see, or rather how they would see were they to
have the eyes of a human being.

Invisible Pin

Stick a pin into a flat round piece of cork and float
it in a bowl with the pin turned downwards. However
you bend your head, you won't be able to see the
pin, though it is long enough for the cork not to hide
it, provided, of course, that the cork is not too big
(Fig. 116). Why do the rays from the pin fail to reach

Fig. 116

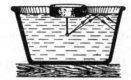

Invisible pin.

your eye? Because .they experience what physicists
call "total internal reflection".

Fig. 117 shows how rays pass from water into air—
or, generally, from a medium with greater refractivity
to one with a smaller refractivity—and back. Passing
from *air into water*, rays bend to draw closer to the
"perpendicular of incidence"—for example, a ray
striking the surface of the water at the angle β to the
perpendicular, to the plane of incidence, bends to
enter the water already at the angle α which is *smaller*
than β. But what happens when the incident ray glides
along the water's surface at almost a right angle to
the perpendicular? It will enter the water at an angle
smaller than a right angle, to wit, 48.5°. It would
not enter the water if the angle is bigger than 48.5°,

Fig. 117

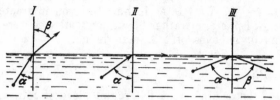

Various instances of refraction when the ray emerges from water into air. In case *II* the ray strikes the surface at the *critical* angle to the normal incidence and emerges from the water to slide along its surface. Case *III* is one of total internal reflection.

since this angle is "critical" for water. These simple relationships have got to be understood so as to comprehend the absolutely unexpected and extremely curious consequences of the laws of refraction that are discussed below.

We have just learned that rays striking the surface of water at various angles are compressed under the water in a rather compact cone with a spread of 48.5+

Fig. 118

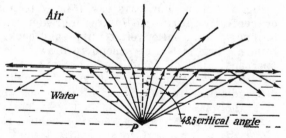

Rays coming from point *P* at an angle to the normal incidence greater than the critical angle (48.5° for water) do not emerge from under the water, but are totally reflected inwards.

+48.5=97°. Let us now see how rays go the other way—*from water into air* (Fig. 118). According to the laws of optics, the rays will follow the same paths. All the rays confined in the afore-mentioned 97°-cone, will emerge at different angles, spreading out along the entire 180° above-water semicircle.

Where then will an underwater ray, lying outside the afore-mentioned cone, go? *It won't emerge at all,* we find, *it will reflect from the surface as from a mir-*

Fig. 119

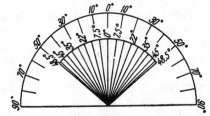

To the underwater observer the 180° arc of the outer world is compressed within a 97° arc, the compression becoming more enhanced the further away the parts of the arc are from zenith (0°).

ror. Generally speaking, every underwater ray striking the surface at an angle greater than the "critical" —more than 48.5°—will not refract but reflect. As the physicist would say, it undergoes a "total internal reflection". (The particular reflection is called a *total* one because all incident rays are reflected. Meanwhile even the best of mirrors, for instance, of polished magnesium or silver, reflect only *part* of the incident rays, absorbing the rest. In the circumstances water represents an ideal mirror.)

For a "physicist fish"—or, perhaps, a "fish-physicist"—the main branch of optics would be the theory of "internal reflection" since it is of prime importance for their underwater vision. Incidentally, the specific

261

features of underwater vision may most likely be associated with the silvery colour of fish scales. Zoologists claim this is due to mimetic adaptation to the water "ceiling" above them. We know that from below, the surface of water is mirror-like due to "total internal reflection". Against this mirror the silvery scales render fish unnoticeable to their larger brethren of the deep that prey upon them.

Underwater Outlook

I am sure many don't even have the slightest inkling of how strange the world would seem were we to take a peep at it from beneath the water. It would be distorted so much that it would be hardly recognisable.

Imagine yourself in this position, getting an underwater perspective. Clouds overhead will appear to you as they always do because perpendicular rays do not refract. However, all other objects, the rays from which strike the surface of the water at acute angles, will be distorted. They will seem much shorter, which is accentuated the more, the more acute the angle of incidence is. And no wonder, since everything spread out in a 180° arc above the water will have to stuff itself into the cramped underwater cone of 97°—nearly half as less—and images are bound to be distorted. All objects from which rays strike the water at the angle of 10° would be so greatly compressed to the view of the underwater observer as to be scarcely distinguishable.

But what would astonish the underwater observer most would be the shape of the water's surface itself. It would appear to be conical, instead of flat. You would think you were at the bottom of a huge crater with sides inclined to each other at an angle a wee bit more obtuse than a right angle—97°. At the top you would see an attractive fringe of all colours of the rainbow. This is due to the different refractive indices, and, consequently, "critical" angles of the various colour components comprising white sunlight.

What would you see beyond this rainbow—haloed fringe? The water's glistening surface itself in which

Fig. 120

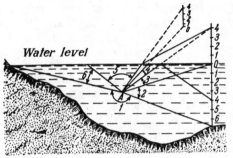

How the underwater observer with eye at point *A* would see a half-submerged depth gauge. Angle *2* shows hazily the submerged part; angle *3* shows its reflection from the inner surface of the water; while still higher you see the jutting part of the gauge contracted and furthermore separated from the rest of the rod. Angle *4* gives a reflection of the bottom. Angle *5* gives the whole of the above-water world in the form of a cone. Angle *6* gives a reflection of the bottom from the lower surface of the water. Angle *1* gives a hazy image of the river-bed.

everything *underneath* would be reflected back as in a mirror.

Incidentally anything half in the water, half out, would present a most fantastic sight to the underwater observer.

Suppose you had a depth gauge sticking out of a river (Fig. 120). What would the underwater observer see at point *A*? Let us divide the observable area of 360° into different sections and discuss each one separately. Within the limits of the angle *1* he would see the river-bed, provided there was enough light. Within the limits of angle *2* he would get an undistorted view of the underwater part of the depth gauge. Roughly within the limits of angle *3*, he would see the reflection of this same part of the gauge, only upside down, due to "total internal reflection". A bit higher our underwater observer would see part of the gauge sticking out of the water, he would see it, however, not as the continuation of the underwater part but much higher and completely separated from its lower half. Indeed, it would never cross his mind that the section of a gauge hovering in the air really belonged to the gauge he saw beneath the water. In addition the gauge would seem greatly compressed especially at the bottom, where its divisions would be noticeably thicker.

To our underwater observer a tree on a bank sub-

How a partly submerged tree would appear to an underwater observer (compare with Fig. 120).

Fig. 122

How a bather up to his chest in the water would appear to an underwater observer (compare with Fig. 120).

merged by spring floods should appear as shown in Fig. 121, while a bather would appear as depicted in Fig. 122. Such is the fish's view of things! When a person is wading in a shallow place, fish see him as two separate split personalities: the top one is legless and the bottom one headless but four-legged. The further away he wades from the underwater observer, the more and more does the top of the body seem compressed until, at some distance, only a floating head remains.

Could we ever test this ourselves? By diving in we would see very little, even if we could get ourselves to keep our eyes open. In the first place the water surface would fail to resume its placid state in those few seconds during which we would be able to stay under water, and it is very hard to make anything out on a *rippling* surface. In the second place, as I have already had occasion to note, the refractivity of water is almost the same as that of the transparent elements of our eye and, consequently, our retinal image would be extremely blurred. Nor would observation from inside a diving bell or helmet or through the porthole of a submarine help. Though the observer will be under water, he will not be in the conditions of underwater vision. I explained earlier that before the rays would reach the eye they would have to *pass again through air*, and would undergo reverse refraction—in which case they would either regain their previous direction or veer off at an angle which would not be the same as in the water. That is why observation through the glass portholes of submarines will never provide the correct idea of what underwater vision is really like.

However, there is no need to dive in *ourselves* to get an underwater "outlook". We can simulate it by using a special water-filled photographic camera, having a metal plate with a hole in the middle, in place of the usual lens. Naturally, if all the space between the aperture and the light-sensitive plate is filled with water, the outer world should appear on the film just as it would to an underwater observer. This, incidentally, was how the American physicist, Professor Wood, took some very curious photographs, one of which is reproduced in Fig. 115. We mentioned this picture earlier and also told you why the straight bridge appeared as an arc.

There is one other way of getting an underwater "perspective" at first hand. This is to submerge a mirror in a lake's placid waters and, after tilting it at the required angle, observe in it the reflection of above-water objects. This will corroborate to the last detail the theoretical points we made above.

To sum up. A translucid layer of water distorts to the eye everything that lies beyond, imparting truly fantastic forms and shapes. Any land creature that would suddenly take to living in water—supposing it could do that—would completely fail to recognise its old habitat, as through the transparent water it would present an entirely different sight.

Underwater Palette

The American biologist Beebe gives an extremely colourful description of changing tints under water.

"At 9:41 in the morning we splashed beneath the surface, and often as I have experienced it, the sudden shift from a golden yellow world to a green one was unexpected. After the foam and bubbles passed from the glass, we were bathed in green; our faces, the tanks, the trays, even the blackened walls were tinged. Yet from the deck, we apparently descended into sheer, deep ultramarine. ...

"... the first plunge erases, to the eye, all the comforting, warm rays of the spectrum. The red and the orange are as if they had never been, and soon the yellow is swallowed up in the green. We cherish all these on the surface of the earth and when they are winnowed out at 100 feet or more, although they are only one-sixth of the visible spectrum, yet, in our mind, all the rest belongs to chill and night and death.

"The green faded imperceptibly as we went down, and at 200 feet it was impossible to say whether the water was greenish-blue or bluish-green. ...

"At 600 feet the colour appeared to be a dark, luminous blue ... it seemed bright, but was so lacking in actual power that it was useless for reading and writing. ...

"I tried to name the water: blackish-blue, **dark**

grey-blue. It is strange that as the blue goes, it is not replaced by violet—the end of the visible spectrum. That has apparently been absorbed. The last hint of blue tapers into a nameless gray, and this finally into black, but from the present level down, the eye falters, and the mind refuses any articulate colour distinction. The sun is defeated and colour has gone forever, until a human at last penetrates and flashes a yellow electric ray into what has been jet black for two billion years."

Elsewhere Beebe says the following about the murkiness at great depths:

"A few days ago the water had appeared blacker at 2,500 feet than could be imagined, yet now to this same imagination it seemed to show as blacker than black. It seemed as if all future nights in the upper world must be considered only relative degrees of twilight. I could never again use the word BLACK with any conviction."

Blind Spot

If I told you that in your field of vision you have a section you don't see at all, though it's right before your nose, you would probably think I was pulling your leg. Indeed, do we never notice in all our life-time such a signal defect? Here is a simple experiment to show that there really is such a thing.

Hold Fig. 123 some 20 cm away from your right eye, meanwhile cupping your left eye, and look at the cross on the left. Then gradually bring the drawing nearer. At one moment the large black spot at the conjunction of the two circles will vanish without trace. You won't see it, though it will still be within the limits of your field of vision and the *two circles at the right and left will be quite distinct*.

This experiment, which the famous physicist Mariotte first performed somewhat differently in 1668, served to amuse the court of Louis XIV. Mariotte would seat two courtiers two metres apart face to face, and ask them to look with one eye closed at a spot placed somewhat to the side. Each thus saw a headless vis-a-vis.

Fig. 123

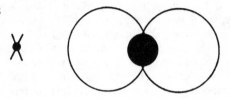

How to find the blind spot.

Strangely enough it was only in the 17th century that man came to know of the blind spot on his retina. This is the area of the retina where the optic nerve enters the eyeball being not yet divided into fine ramifications, and where the retina has no light-sensitive elements.

We never notice this black "hole" in our vision because we have long grown accustomed to it and because our imagination involuntarily makes up for the deficiency. Thus, though we don't see the spot in Fig. 123, we continue the lines mentally and are quite sure that we clearly see their intersection.

Do you wear glasses? Then you might like to try the following experiment. Glue a scrap of paper to one of the lenses—only not quite dead centre. For a few days it will be a nuisance, but after a week or two you will cease to notice it at all. Incidentally, if you ever happen to crack your glasses and wear them for some time after, you might remember that the crack annoyed you only at the beginning. Habit, and only habit, is again responsible for our being "blind" to our blind spot. Then one must note that the blind spots of the right eye and the left eye correspond to different places in the two fields of vision. Consequently, with two-eyed vision, there is no deficiency in the range they both cover.

Don't think the blind spot is nothing to "write home about". Whenever you look, with one eye closed naturally, at a house, say, some 10 metres away, you fail, on account of the blind spot, to see quite a big portion of its façade of more than a metre across—enough to accommodate a whole window. While if you turned your eye to the heavens you would fail to see a space equal in area to 120 full moons.

How Big We Think the Moon

A few words, by the by, about the moon's visible proportions. Ask your friends how big they think the moon is. You're likely to get a host of different replies. Most will say the moon is as large as a plate, but some might think it the size of a saucer, an apple, or even a cherry. A schoolboy I once knew always thought the moon "as big as a round table covered for twelve" while a certain writer has claimed in a book that the moon is a "yard across".

Why do we differ so much with regard to the size of one and the same thing? Because we *estimate distances* differently and moreover subconsciously. A person who takes the moon to be as large as an apple, imagines it to be much nearer than people who think it the size of a plate or a round table.

I noticed before that most think the moon as large as a plate and thereby hangs a tale. If we compute the distance—later I shall tell you how this is done—

Fig. 124

When looking at the building with one eye, the small section c' of the field of vision, corresponding to the *blind spot* c of the eye, is not seen at all.

at which we put the moon with such *apparent* dimensions, we shall find that our distance is not more than 30 metres! (See Minaert's *Light and Colour in Nature* for greater detail about this and other allied matters.) Such is the very modest distance to which we subconsciously remove our nocturnal luminary.

Incidentally, quite a few optical illusions are based on the erroneous estimation of distance. I well remember one to which I succumbed when I was a boy. One spring I went out of town on an outing for the first time in my life. I saw a herd of cows browsing in the meadow. They appeared as pygmies to me; I had misestimated the distance. I have never seen such tiny cows ever since and, naturally, never will. (Adults also succumb to this illusion—as borne out by the following extract form the story "Ploughman" by the

Fig. 125

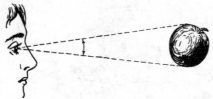

19th-century Russian writer Grigorovich: "The country seemed to fit in the hollow of one's hand. The trees seemed fast by the bridge; while the cottage, hill and birch thicket now seemed adjacent to the village. Everything—house and orchard and village—resembled playthings, with stalks for trees and the slivers of a mirror for a river.")

Astronomers define the apparent dimensions of celestial bodies by taking the angle at which they are seen. The "angle of vision" is the angle between two straight lines traced to the eye from the extremes of the observed object (Fig. 125). As you all very well know, angles are measured in degrees, minutes, and seconds. When asked about the moon's apparent dimensions, the astronomer won't say that it is as large as an apple or plate. He will say "half a degree", which means that there is an angle of half a degree between the straight lines from either limb of the lunar disc to our eye. This definition of apparent dimensions is the only correct one as it does not produce misunderstanding.

Geometry teaches us that an object, set away from the eye at a distance 57 times greater than its diameter, is observed at an angle of 1°. For instance, an apple five centimetres in diameter will have an angular value of 1° if held at a distance of 5×57 cm away. At twice that distance it would appear at an angle of half a degree or as large as we see the moon. You can even say that the moon seems to be as large as an apple—provided the latter is 570 cm away. Should you want to compare the moon's apparent dimensions to a plate, you will have to move the plate some 30 metres away. Most are loth to believe that the moon seems so small. Well, place a six-penny bit at a distance 114 times greater than its diameter. Though this will be but two metres, it will cover the moon.

If I told you to draw a circle for the lunar disc as seen by the naked eye, you would most likely think the task not concrete enough, as your circle could be big or small—depending on how far it was from the eye. Let us then choose the distance at which we usually hold a book or drawing—that is, the distance of optimal vision, which for the normal eye is 25 cm. Let us see how big a circle, on a page of this book, for example, should be, for its apparent size to equal the moon's disc. The problem is simple enough; all we need do is divide the 25-cm distance by 114. Our result will be tiny, a wee bit more than 2 mm—about the same size as the letter "o" in this book. It's really hard to believe that the moon as well as the sun—which is of equivalent apparent dimensions—seem so small at such a small angle.

273 You have probably noticed that after you look at

the sun for a while, you seem to see for quite a time a jumble of coloured discs. These so-called "optical traces" have the same angular value as the sun but their apparent dimensions change. When you look at the sky, they are as large as the sun itself, but as soon as you look down at the book, the sun's "trace" on the page will occupy a space of no more than the letter "o", which graphically demonstrates the correctness of our calculations.

Apparent Dimensions of Celestial Bodies

If we were to draw, keeping to angular dimensions, the constellation of Ursa Major, the Dipper, we would get the figure depicted in Fig. 126. Looking at it

Fig. 126

Dipper constellation with angular dimensions preserved. This figure should be held 25 cm away from the eye.

from the distance of optimal vision we would see it as it appears to us in the sky. This, so to speak, would be a map of the constellation with angular dimensions preserved. If you are familiar with the visual impression this constellation produces—that is not only its *pattern*, but namely the direct *visual impression*—you will get this same impression when looking at the drawing appended. Knowing the angular distance between the main stars of all the constellations—they are given in astronomical calendars and ephemerides—you will be able to give "full-size" drawings for a whole astronomical atlas. To do this you must have some millimetre lined paper and reckon each 4.5 mm as one degree (the areas of the circles representing the stars should be given in proportion to their brilliance).

Let us now turn to the planets. Their apparent dimensions—as of the stars—are so tiny that to unaided vision they seem radiant points of light. No wonder, because no planet, with the exception, perhaps, of Venus during the period of its greatest brilliance, ever appears to the naked eye at an angle of more than 1′, which is the critical value at which we will be able generally to distinguish an object as a dimensional body (at a smaller angle every object seems just a point).

Here is a table of the dimensions in angular seconds of the different planets; the two figures opposite each planet denote respectively the distances when the planet is closest to, and furthest from, the earth.

Planet	Seconds
Mercury	13-5
Venus	64-10
Mars	25-3.5
Jupiter	50-31
Saturn	20-15
Saturn's rings	48-35

It is impossible to give these dimensions in their "full size". Even a whole angular minute of 60° corresponds, at the distance of optimal vision, to but 0.04 mm, which is too small for the normal eye to see. We shall therefore depict the planetary discs as seen through a telescope with a magnifying power of 100.

Fig. 127

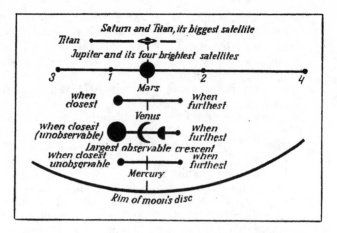

Hold this diagram 25 cm away from the eye and you will see the planet discs as seen in a telescope with a magnifying power of 100.

Fig. 127 gives the apparent dimensions of the planets thus magnified. The lower arc is the rim of the moon's or sun's disc as seen through a 100-power telescope. Right above it is Mercury when closest to the earth. Still higher are the different phases of Venus. At its most favourable opposition it is not seen at all as it turns its unlit face towards the earth. (It is seen in this phase only during those extremely rare periods when projected on to the sun's disc as a black circle — the so-called "passage of Venus".) Then comes its visible slender crescent which is biggest of all planetary "discs". In subsequent phases Venus grows smaller and smaller till it becomes a full disc with a diameter only a sixth of that of the crescent.

On top of Venus is Mars, shown on the left as when closest to the earth and on the right as when furthest away. The left-hand disc, let me remind you, is what Mars looks like in a telescope with a magnifying power of 100. Do you think anything could be made out on so small a disc? Imagine it 10 times bigger and you will have some idea of what an astronomer sees when he studies Mars in a very powerful telescope with a magnifying power of 1,000. Again, I ask you, do you think it possible to make out with unquestionable clarity on so small an object such details as the notorious canals or to detect the slight variations in coloration supposedly due to vegetation on the bed of the "seas" of this strange world? No wonder some observers essentially contradict what others seem to see. Some regard as an optical illusion what others claim to discern quite distinctly.*

The giant planet of Jupiter and its satellites occupy an extremely prominent place in our table. This planet's disc is much greater than those of all the other planets—with the exception of Venus' crescent—while its four main satellites are spread out along a line equal to nearly half the moon's disc.

In the figure Jupiter is shown as when closest to the earth. Finally comes Saturn with its rings and its biggest moon Titan—which are also rather noticeable in most favourable opposition.

After all I have said, it should be clear to you that every visible object appears to be the smaller the closer we imagine it. Conversely, if, for some reason, we would exaggerate the distance to the object, the object itself would become correspondingly larger.

---

* Present available information on Mars and other planets is not limited exclusively to data obtained by visual observation alone. Thanks to measurements performed by means of sensitive instruments, we have been able to draw some quite definite and veracious conclusions as to the physical conditions on the planets and their satellites.—*Ed.*

Let me now regale you with an instructive story by Edgar Allan Poe describing precisely one such optical illusion. Though it seems scarcely true it is not at all a figment of the imagination. I myself succumbed once to practically the same type of illusion and many of you may recall a similar experience.

The Sphinx

By Edgar Allan Poe

(The story is given with a few minor cuts)

"During the dead reign of cholera in New York, I had accepted the invitation of a relative to spend a fortnight with him in the retirement of his *Cottage ornée.* ... We should have passed the time pleasantly enough, but for the fearful intelligence which reached us every morning from the populous city. Not a day elapsed which did not bring us news of the decease of some acquaintance. At length we trembled at the approach of every messenger. The very air from the South seemed to us redolent with death. That palsying thought, indeed, took entire possession of my soul. ... My host was of a less excitable temperament, and, although greatly depressed in spirit, exerted himself to sustain my own. ...

"Near the close of an exceedingly warm day, I was sitting, book in hand, at an open window, commanding through a long vista of the river banks a view of a distant hill. ... My thoughts had been long wandering from the volume before me to the gloom and desolation of the neighbouring city. Uplifting my eyes from the page, they fell upon the naked face of the hill and upon an object—upon some living monster of hideous conformation, which very rapidly made its way from the summit to the bottom, disappearing finally in the dense forest below. As this creature first came in sight, I doubted my own sanity—or at least the evidence of my own eyes—and many minutes passed before I succeeded in convincing myself that I was neither mad nor in a dream. Yet when I describe the monster (which I distinctly saw, and calmly surveyed through the whole period of its progress), my readers, I fear, will feel more difficulty in being convinced of these points than even I did myself.

"Estimating the size of the creature by comparison with the diameter of the large trees near which it passed ... I concluded it to be far larger than any ship of the line in existence. I say ship of the line, because the shape of the monster suggested the idea— the hull of one of our seventy-fours might convey a very tolerable conception of the general outline. The mouth of the animal was situated at the extremity of a proboscis some sixty or seventy feet in length, and about as thick as the body of an ordinary elephant. Near the root of this trunk was an immense quantity of black shaggy hair ... and projecting from this hair downwardly and laterally, sprang two gleaming tusks not unlike those of the wild boar, but of infinitely greater dimension. Extending forward, parallel with the proboscis, and on each side of it, was a gigantic staff, thirty or forty feet in length, formed seemingly of pure crystal, and in shape of a perfect prism—it reflected in the most gorgeous manner the rays of the declining sun. The trunk was fashioned like a wedge with the apex to the earth. From it there were outspread two pairs of wings—each wing nearly one hundred yards in length—one pair being placed above the other, and all thickly covered with metal scales; each scale apparently some ten or twelve feet in diameter.... But the chief peculiarity of this horrible thing was the representation of a *Death's Head*, which covered nearly the whole surface of its breast, and which was as accurately traced in glaring white, upon the dark ground of the body, as if it had been there carefully designed by an artist. While I regarded this terrific animal, and more especially the appearance on its breast, with a feeling of horror and awe ... I perceived the huge jaws at the extremity of the proboscis suddenly expand themselves, and from them there proceeded a sound so loud and so expressive of woe, that it struck upon my nerves like a knell, and as the monster disappeared at the foot of the hill, I fell at once, fainting, to the floor.

"Upon recovering, my first impulse, of course, was to inform my friend of what I had seen and heard.... He heard me to the end—at first laughed heartily— and then lapsed into an excessively grave demeanour, as if my insanity was a thing beyond suspicion. At this instant I again had a distinct view of the monster —to which, with a shout of absolute terror, I now directed his attention. He looked eagerly—but maintained that he saw nothing—although I designated minutely the course of the creature, as it made its way down the naked face of the hill.... I threw myself passionately back in my chair, and for some moments buried my face in my hands. When I uncovered my eyes, the apparition was no longer visible.

"My host ... questioned me very vigorously in respect to the conformation of the visionary creature. When I had fully satisfied him on this head, he sighed deeply, as if relieved of some intolerable burden ... stepped to a book-case, and brought forth one of the ordinary synopses of Natural History. Requesting me then to exchange seats with him, that he might the better distinguish the fine print of the volume, he took my arm-chair at the window, and opening the book, resumed his discourse very much in the same tone as before.

"'But for your exceeding minuteness,' he said, 'in describing the monster, I might never have had it in my power to demonstrate to you what it was. In the first place, let me read to you a schoolboy account of the genus *Sphinx*, of the family *Crepuscularia*, of the order *Lapidoptera*, of the class of *Insecta*—or insects. The account runs thus:

"'Four membranous wings covered with little coloured scales of metallic appearance; mouth forming a rolled proboscis, produced by an elongation of the jaws, upon the sides of which are found the rudiments of manibles and downy palpi; the inferior wings retained to the superior by a stiff hair, antennae in the form of an elongated club, prismatic; abdomen pointed. The Death's-headed Sphinx has occasioned much terror among the vulgar, at times, by the melancholy kind of cry which it utters, and the insignia of death which it wears upon its corslet.'

"He here closed the book and leaned forward in the chair, placing himself accurately in the position which I occupied at the moment of beholding the monster.

"'Ah, here it is,' he presently exclaimed—'it is reascending the face of the hill, and a very remarkable looking creature I admit it to be. Still, it is by no means so large or so distant as you imagined it; for the fact is ... it wriggles its way up this thread, which some spider has wrought along the window-sash....'" (This butterfly has now been classified as a member of the *Acherontia* family. It is one of the few butterflies able to emit sounds—in this case a whistle resembling the squeaking of a mouse—and the only butterfly to do so with its oral organs. The sound is rather shrill and can be heard many metres away. In this particular case, it may have seemed very loud, as the observer had thought its source to be rather far away—see Chapter Ten "The Tricks Our Ears Play" of Book One of *Physics for Entertainment*.)

Why Does a Microscope Magnify?

"Because it alters the direction of the rays in a definite way as is described in textbooks on physics," is the reply one hears most often. This however is but a remote reason, one having no relation to the real cause. So what really causes the microscope and telescope to magnify?

I learned the answer accidentally, when a schoolboy, and moreover not from any textbook. I happened to notice an extremely curious and perplexing thing. I was sitting near a closed window looking at the brick wall of a house across the road. Suddenly I recoiled in horror. I could clearly see a huge eye several metres wide staring at me from the wall. I hadn't read Edgar Allan Poe's story then and did not realise at once that this huge eye was a reflection of my own eye, a reflection I had projected on this wall and had therefore imagined as correspondingly enlarged.

When I finally realised what had happened, I started wondering whether a microscope based on the same optical illusion couldn't be made. My efforts were abortive, but one good thing ensued—I realised why a microscope magnifies. This is not at all because the object viewed appears to be larger, but because it is viewed at a *much wider angle of vision*, with the result—and this is the most important thing—*that its image occupies a far greater retinal area*.

To help you understand why the angle of vision is so essential, let me refer to a significant feature of our eye. Every object or part of an object, that we observe at an angle of less than one minute, appears to the normal eye as a *point* with no distinguishable shape or elements. When an object is far away or so small that the whole of it, or parts of it, are observed at an angle of vision less than one minute, we cease to make out its details. This is because at this angle of vision the retinal image of the object or its element covers only one visual cell, with the result that we see a *point* and nothing more.

The microscope and telescope alter the direction the rays from the object under observation take, and thus present it at a greater angle of vision. The retinal image is stretched out to cover more visual cells and we are thus able to distinguish details which had previously merged into one dot. When people say that a microscope or telescope has a magnifying power of 100, they mean that it presents objects at an angle of vision 100 times greater than the angle at which we would see them without the instrument. If an optical instrument does not increase the angle of vision, *it will not magnify at all, even if we think we are seeing a magnified object.* The eye on the brick wall seemed tremendous, but I did not see *any new detail* in addition to what I saw when looking in a mirror. When low on the horizon, the moon seems much *larger* than high up in heavens; but we fail to detect on this enlarged disc anything new which we didn't notice when our nocturnal luminary was high in the sky.

Taking the case of magnification Edgar Allan Poe describes, we shall again see that no new details were observed in the magnified object. The angle of vision remains constant, and the butterfly is observed at one and the same angle whether removed further away or brought up close to the window. Since the angle of vision does not change, the enlarged object, however staggering, doesn't reveal a single new detail. A faithful realist Edgar Allan Poe remains loyal to nature even on this point. I wonder if you noticed how he describes the "monster" in the wood. The list of the insect's different members does not add anything new to what this "dead head" had when seen with the unaided eye. Compare the two descriptions—given not without reason—and you will see that they differ only in the metaphors used (ten-foot scales—little scales,

Fig. 128

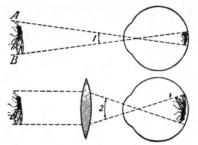

The lens increases the retinal image.

giant tusks, stiff hair, etc); there are no new details.

If this were the only way a microscope magnified, scientists would find it useless; it would be nothing more than a curious plaything. However we know this not to be so; we know that the microscope opened up a new world, extending the boundaries of our natural vision.

Now we can explain why the microscope gives us details which Edgar Allan Poe's observer failed to see in his monster butterfly. The microscope gives not just an enlarged image and nothing more; it presents the object *at a greater angle of vision*. Hence the retinal image is *larger*, covering a bigger number of visual cells and thus supplying a bigger number of individual visual impressions. In short, the microscope magnifies not the object but its retinal image.

Visual Self-Deception

We often use the term "optical illusion" or "auditory illusion". However there can be no deception of the *senses*. The philosopher Kant aptly noted in this connection: "The senses do not deceive us. Not because they always rightly pass judgement, but because they don't pass judgement at all."

What then deceives us? Naturally, what passes *judgement* in this case—our bram. Indeed, most op-

Fig. 129

Which is wider, the right-hang figure or the left-hand one?

tical illusions derive not only from what we *see* but from what we subconsciously *think* we see; we involuntarily deceive ourselves. This is a deception produced by what we *think* we see or hear, not by what the senses actually register.

Some two thousand years ago Lucretius wrote:

"Our eyes are unable to take cognition of the nature of objects, so do not impose upon them delusions of the mind."

Let us take a commonly known case of an optical illusion. The picture on the left in Fig. 129 seems narrower than that on the right, though both are limited by identical squares. This is because in estimating the *height* of the picture on the left we subconsciously add

285

the gaps between the lines. Therefore it seems to be higher, though actually both are the same. For the same reason the figure on the right seems to have more width than height, and again for the same reason. Fig. 130 seems to have more height than width.

Fig. 130

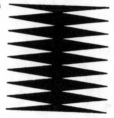

Which is larger, height or width?

Illusions Useful for Tailors

When you try to apply the optical illusion just described to bigger figures which the eye is unable to take in at one glance, your expectations fail to materialise. You all know that a small fat man in a pin-striped suit in which the stripes are horizontal, seems still fatter. On the contrary, by donning a pin-striped suit in which the stripes are vertical, a fat man can make himself appear a bit slimmer.

Why the contradiction? Because we are unable to take in the entire suit at one glance, without moving the eye; as we do that, we involuntarily let our eye travel down the stripes. The effort our eye muscles make in this process compels us subconsciously to magnify the object in the direction the stripes go, as we are accustomed to associating this exertion with notions of larger objects that do not lie completely within the field of vision. On the other hand, when we look at a *small* striped drawing, our eye doesn't move and the muscle does not have to exert itself.

286

Which Is Bigger?

Which ellipse in Fig. 131 is bigger—the one at the bottom or the inner one on top? The bottom one really seems bigger than that at the top, doesn't it? Actually both are identical. It is only because of the outer

Fig. 131

Which ellipse is bigger, the bottom one or the top inner one?

fringing ellipse that we get the illusion that the inside ellipse is smaller than the bottom one. This illusion is enhanced furthermore by the entire figure appearing to be not flat but three-dimensional, and shaped as a pail. We involuntarily mentally redraw ellipses as circles compressed into perspective and the straight side lines as slanting ones standing for the sides of the pail.

In Fig. 132 the distance between points *a* and *b*

Fig. 132

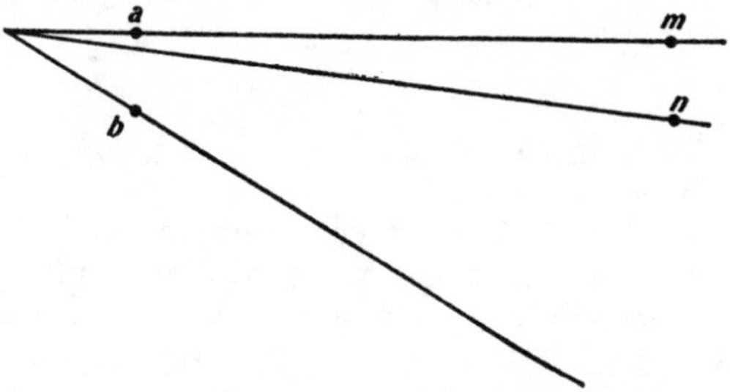

Which distance is longer, *ab* or *mn*?

seems bigger than that between points *m* and *n*. The presence of a third straight line, commencing from the same apex, enhances the illusion.

The Power of Imagination

As I have already noticed, most optical illusions depend not only on what we see but also on what we *imagine* we see. "We look with our brain and not with our eyes," physiologists say. You will surely agree with this after an acquaintance with several il-

Fig. 133

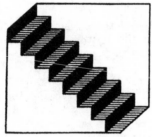

Fig. 134

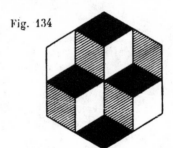

What is this, a staircase, a niche, or a strip of paper creased like a concertina?

How are the cubes disposed here? Where do you see two cubes side by side, on top or at the bottom?

lusions in which imagination plays a *conscious* role in the process of vision.

Look at Fig. 133. If you were to show it to your friends, you might get three different answers as to what it represents. One might say it is a staircase, another, that it is a niche in a wall, and a third, that it is a strip of paper creased like a concertina and stretched across a white square.

Oddly enough all three answers are right. You yourself will get all three illusions, provided you look at it differently. First, cast your glance at the *left* half of the drawing—you will see a staircase. Then let your eye travel across the drawing from right to left— you will see a niche. Finally, if you let your eye travel slantwise, diagonally from the bottom right-hand

Fig. 135

What is longer, *AB* or *AC*?

corner to the top left-hand corner, you will see a strip of paper creased like a concertina. Incidentally, when you look too long, your attention will stray and you will get alternately the first, second or third illusion, regardless of what you would really want to see.

Fig. 134 presents a similar case.

The illusion Fig. 135 produces is most curious. We involuntarily succumb to the impression that the distance *AB* is shorter than *AC*, though actually they are the same.

More Optical Illusions

We are unable to explain all the optical illusions. We often cannot guess what subconscious conclusions our brain makes to cause one or another optical illusion. Fig. 136 distinctly shows two arcs, their bulges

Fig. 136

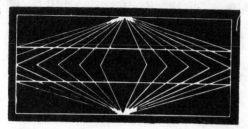

The two lines in the middle going from right to left are straight parallel lines, though they seem to be arcs with facing bulges. The illusion vanishes when (1) you raise the figure to eye level and look along the lines, or when (2) you place a pencil tip at any point in the figure and concentrate on this point.

Fig. 137

Is this straight line divided into six equal parts?

facing each other. You hardly believe it, when, by putting a ruler to these "arcs" or looking at them lengthwise with the figure at eye level, you find that they are straight. Why the illusion? That is not so easy to explain.

Some more instances of similar illusions. In Fig. 137 the straight line seems divided into unequal portions. Measure them and you find they are equal. In

290

Fig. 138

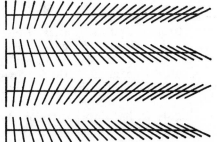

The parallel lines in this figure don't seem to be parallel.

Fig. 139

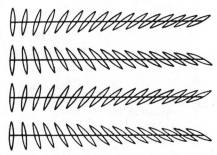

An illusion similar to the previous figure.

Fig. 140

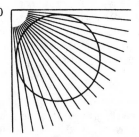

Is this a circle?

Figs. 138 and 139 the straight parallel lines seem bent. In Fig. 140 the circle seems oval.

It is curious to note that the optical illusions provided in Figs. 137, 138 and 139 cease to deceive if looked at by the light of an electric spark. The illusions are evidently associated with a movement of the eye—which is too slow for the spark's instantaneous flash.

One more highly entertaining illusion known as "the pipe". Look at Fig. 141 and tell me which of the

Fig. 141

The pipe illusion. The strokes on the right seem shorter than those on the left, though actually they are identical.

lines are longer—those on the left or those on the right? The left lines seem longer though actually both are the same (this drawing, incidentally, illustrates the well-known geometrical principle of Cavalieri stating that the areas of the two parts of the "pipe" are equal). Many reasons have been suggested to explain these curious illusions, but they are not very convincing and I shan't bother to adduce them. However, one thing is plain, the illusions derive from the subconscious mind which prevents us from seeing what we really should see.

What Is This?

I doubt whether you would guess at once what Fig. 142 depicts. You would probably dismiss it as a circular piece of black mesh. However, stand the book upright on the table, take some three or four steps back and look again. You see a human eye. Immediately you draw up nearer, you again see nothing but the mesh. You might think it an artful trick by an ingenious engraver. No, this is but a crude instance of the optical illusion we succumb to every time we

Fig. 142

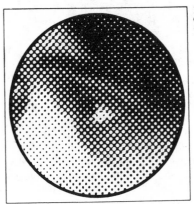

From a distance you see an eye and part of the nose of a woman's face turned to the left.

look at half-tone reproductions. Book and magazine reproductions always seem to present a solid background, but when you peer at them through a magnifying glass you see a similar pattern of dots as shown in Fig. 142. The brain-teaser is part of an ordinary book illustration magnified ten times over. When the gauze is fine it fuses into one solid background already close up, at the distance we usually hold a book when reading. But when the gauze is coarse, it will·blend into one solid background only at a much greater distance away. You need only remember all that I said about the angle of vision to realise why this is so.

Have you ever watched through a crack in a fence or, still better, on the movie screen the wheel spokes of a quickly moving cart or motor-car? You've most likely noticed a strange circumstance: the motor-car rolls along, but its wheels hardly move, if at all. Sometimes you see them spinning in an opposite direction. This optical illusion is so unusual that it perplexes all who see it for the first time.

Here is the explanation. When watching a spinning wheel through a crack in a fence, we see its spokes not continuously but in a series of regular identical time intervals—since the fence shuts them to view every instant. In the same way will a movie film depict wheel images intermittently, 24 frames a second. This gives rise to *three different cases*, which we shall now proceed to discuss one after the other.

In the first case, during the first time interval, the wheel makes any number of *full revolutions*— whether two or twenty it makes no difference. In this case, the wheel spokes will assume in the next frame the same position as in the previous one. In the next time interval the wheel again makes the same number of *full revolutions* (the length of the time interval and the speed of the motor-car have remained constant), and the spokes remain in the same position. As we see the spokes in one and the same position we conclude that the wheel is not turning at all (middle column in Fig. 143).

In the second case, the wheel makes during every time interval a number of full revolutions plus a *fraction of a revolution*. As we watch the series of images we see not the whole number of revolutions made, but merely a gradual turning—by a small fraction of a revolution each time. As a result we conclude that

Fig. 143

True direction of rotation

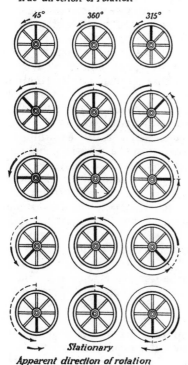

Apparent direction of rotation

the wheel is turning slowly, though the motor-car itself may be moving very quickly.

In the third case, the wheel makes an incomplete revolution during the time interval between two frames—just a small fraction less than a complete revolution (315°, for instance, as shown in the third column in Fig. 143). In this case the wheel spoke appears to be turning in the *opposite direction*—an illusion which lasts until the wheel goes faster or slower.

There remains but little to be added. In the first case, for the sake of simplicity, we gave a number of full revolutions; however, since wheel spokes look all alike, it is quite enough for the wheel to turn a whole number of *segments*, between the spokes. The same holds for the other two cases as well.

Then there are, furthermore, the following interesting points. If the wheel rim is marked at some point, we may see the rim moving in one direction and the spokes—which are all alike—in the opposite direction. Should a spoke be marked, we may see the spokes moving in a direction contrary to the mark which will seem to be skipping from spoke to spoke.

When you happen to be watching a feature film or a newsreel, this illusion does not annoy you. But if the film-maker seeks to explain a machine's principle of operation, this optical illusion may lead to serious misunderstandings and even provide a totally distorted picture of how the machine really works.

The attentive movie-goer will easily be able to figure out how many revolutions a wheel is making a second, by counting its spokes. The usual speed of film projection is 24 frames a second. In the case of a twelve-spoke wheel, the number of revolutions per second will be equal to $24:12=2$, or one whole revolution every half-second. This is the least number of revolutions; it may be greater by any whole number of times—twice as much, three times as much, and so on. By estimating the wheel's diameter we may even try to guess how fast the motor-car is moving. For a wheel diameter of 80 cm, in the case just described, the motor-car's speed would average 18 km/h—or 36, 54, etc., km/h.

Engineers avail themselves of this optical illusion to count the number of revolutions of quickly rotating shafts. This is how they do it. The luminous power of a bulb consuming a-c current does not stay constant. Every one-hundredth of a second the light dims, though usually we don't see the twinkle. However, imagine that the revolving disc depicted in Fig. 144 is lit by such a bulb. Supposing the disc were to make 1/4 of a revolution every one-hundredth of a second, we would see, instead of the uniformly grey circle, the black and white segments, as if the disc were sta-

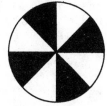

Fig. 144

tionary. All who have understood the causes of the
wheel illusion will also realise why this happens and
how it can be used to count the revolutions of a rotating shaft.

"Slow-Motion Microscope" in Technology

In Book One of *Physics for Entertainment* I told
you about the slow-motion camera. There is another
way this effect can be achieved on the basis of the phenomenon I have just described. You already know that
when the type of disc shown in Fig. 144 and making 25
revolutions a second is lit every second by 100 flashes
it seems stationary. Imagine now that we have 101
flashes a second instead of 100. In the intervals between two successive flashes, the disc will no longer
turn a full quarter of a revolution as it did before and,
consequently, the corresponding sector will not reach
its initial position. One will see it lag behind by a
hundredth of the circumference. In the next flash it
will seem to lag still further behind by another hundredth and so on and so forth. The disc will appear
to be turning *backward* making one full revolution a
second. Motion will have thus been slowed down 25
times.

It is not so hard to guess how one could see the same
sort of slow-motion rotation, but going forward and

not backward. To do that we must *reduce*, instead of increase, the number of flashes. For instance, with 99 flashes a second, the disc would seem to be going forward, making one revolution a second.

This is what we could call a "slow-motion microscope" with a time-delaying power of 25. We could slow down motion still more. If, for example, we had 999

Fig. 145

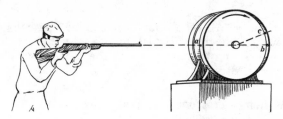

How to gauge the speed of a bullet in flight.

flashes every 10 seconds or 99.9 per second, our disc would seem to be making one revolution every ten seconds, and we would have achieved a time-delaying power of 250.

Any fast, periodically recurring, motion can thus be slowed down to the required degree. This provides a handy method for studying the specific features of the motion of very fast mechanisms*.

In conclusion, I shall describe a method employed to measure the speed of a bullet in flight. It is based on the possibility of determining the exact number of revolutions made by a revolving disc. A cardboard drum (Fig. 145) is mounted on a quickly spinning shaft. The mark-

---

* The principle just discussed is that of the stroboscope which is used to measure the frequency of rapidly varying processes. It gives exceptional accuracy—for instance, an electronic stroboscope may be only 0.001 per cent out.—*Ed.*

sman fires a bullet and pierces the drum's walls in
two places along its diameter. If the drum were stationary the bullet holes would be at the two ends of one
and the same diameter. But as the drum was turning,
in the time the bullet passed from end to end, it struck
point *c* instead of point *b*. Knowing the number of
revolutions the drum makes and its diameter, we may
compute the bullet's speed on the basis of the value
of the arc *bc*. This is a geometrical problem which is
more or less simple for all possessing some knowledge
of mathematics.

The Nipkov Disc

The so-called Nipkov disc, used in the first televi-
sion sets, was a splendid case of the technical application of an optical illusion. Fig. 146 shows a disc with

Fig. 146   Fig. 147

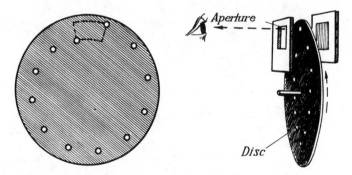

a series of a dozen or so small holes 2 mm in diameter. Set uniformly along a spiral near the outer edge,
each hole is 2 mm closer to the centre than its preced-
ing neighbour. Now you might think there is nothing

remarkable about it. However, mount it on a shaft, place a small window in front of the disc and a picture of the same size behind (Fig. 147), and spin the disc rapidly. The picture, which the disc had shut to view before, can now be seen quite clearly. As soon as the disc slows down the picture becomes blurred before it finally vanishes altogether when the disc stops. All you see of the picture now is what you can make out through the tiny, 2-mm hole.

What is the secret of this mysterious disc? Let us revolve the disc slowly and watch each successive hole as it passes the window. The hole furthest from the

Fig. 148

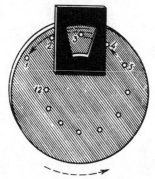

centre passes near the window's upper rim. When moving quickly enough, it opens to view a whole strip along the upper edge of the picture. The next hole, which is a bit below the first, opens, in fast passage, a second strip abutting on the first (Fig. 148), the third hole opens to view a third strip abutting on the second, and so on and so forth. When the disc is revolved rapidly enough, we see the entire picture. Opposite our window there seems to be in the disc itself another aperture of the same size.

This disc is quite easy to make. To spin it you could use a piece of string wound round its shaft, but, of course, it would be better to work it with a small electric motor.

Why Is a Rabbit Cross-Eyed?

Man is one of the few creatures whose eyes are accommodated to the simultaneous perception of an object. It is only by a wee bit that the right eye's

Fig. 149   Fig. 150

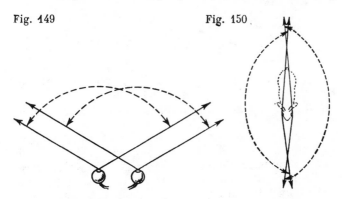

Field of vision of a man's eyes.

Field of vision of a rabbit's eyes.

field of vision does not coincide with that of the left eye. Most animals, however, look with each eye separately. What they see lacks the relief to which we are accustomed, but this is made up for by a field of vision which is much wider than ours.

Fig. 149 shows man's field of vision. Each eye sees —horizontally—within an angle of 120°, and moreover, both angles almost completely overlap, assuming that the eyes are not moving, of course. Now compare this drawing with Fig. 150 which gives the rabbit's field of vision. Without turning its head, the rabbit sees with its wide-set eyes not only what goes on in front but also behind it. The fields of vision of both eyes link up, both in front and behind. This explains why it is so hard to creep up on a rabbit from

Fig. 151

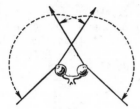

Field of vision of a horse's eyes.

behind without scaring it away. On the other hand, from the drawing it is clear that the rabbit can't see at all what is right in front of its nose without turning its head sideways.

Nearly all hooved and ruminant animals possess this property of "all-round vision". Fig. 151 shows the horse's fields of vision. Though they do not converge at the back, it is enough for the horse to turn its head but slightly to see everything behind it. Visual perceptions, true, are not so distinct, but, on the other hand, the animal will not miss the slightest motion. Agile beasts of prey instead of this "all-round" vision, have a "two-eyed" vision, which enables them accurately to estimate the pouncing distance.

Why Are All Cats Grey When the Candles Are out?

A physicist would say that when the candles are out all cats are black, as in the absence of light nothing is to be seen in general. The proverb does not mean total darkness; rather does it imply that when light is poor we cease to distinguish colour and everything seems grey to us.

Is this really true? Is it really so that in half darkness a red flag and a green leaf will both seem grey? You can easily verify this. If you have ever tried to make out colours in twilight, you must have observed the general more or less darkish grey tinge that every object acquires, regardless of its colour—be it a red blanket, blue wall paper, violet flowers, or green leaves. "The drawn blinds barred the sunshine," we read in Chekhov's story *The Letter*, "and it was semidark. All the roses in the big bouquet seemed of one colour."

Exact physical experiments have quite confirmed this observation. If you turn a faint white light on a coloured surface or a faint tinted light on a white surface, and gradually increase the candle-power, you will see at first only grey and no other colour. Only when the candle power is intensive enough do you begin to notice colour. This is called the absolute threshold of colour perception.

So the same proverb which exists in many tongues is literally right, because below the threshold of colour perception everything seems grey.

There is also a terminal threshold of colour sensation. When light is too bright, the eye again ceases to distinguish colour and all coloured surfaces seem *white*.

Are There Rays of Cold?

The opinion is current that along with rays carrying heat there are also rays that carry cold. After all a lump of ice seems to shed cold in the same way a stove radiates heat. Does not this suggest that ice radiates rays of cold just as the stove radiates rays of warmth?

This is absolutely wrong. Rays of cold simply don't exist. Articles placed next to ice grow cold, not because there are "rays of cold" but because warm articles radiate more warmth than they themselves receive from the ice. Both the warm article and the cold ice radiate heat. An article warmer than ice sheds more heat than it receives. Since the amount of incoming heat is less than the amount of outgoing heat, the article cools.

There is an effective experiment which may induce us to think that rays of cold really exist. Large concave mirrors are hung opposite each other in a long hall. If we place a powerful source of heat in the focus of one of the mirrors, the rays it emits will be reflected from this mirror to the second mirror, which will again focus them; a piece of black paper at this focus will burst into flame. This graphically demonstrates the existence of heat-carrying rays. However if we place a lump of ice at the focus of the first mirror a thermometer at the focus of the second mirror will register a drop in temperature. Does this really mean that the ice emits rays of cold which the mirror reflects and focuses on the thermometer?

Not at all. Again an explanation can be found which will debunk these mysterious "rays of cold". Due to radiation the thermometer will impart to the ice more warmth than was received from it; this is why the temperature drops. Again we have no reason to believe that rays of cold exist. There is nothing of the sort in nature. All rays impart energy to the absorbing object. Meanwhile objects that emit rays grow colder.

10 *Sound. Wave Motion*

Sound and Radio Waves

Sound travels with a speed a million times slower than that of light. Since the speed with which radio waves propagate coincides with that of light oscillations, hence sound is a million times slower than a radio signal. This leads to a curious conclusion, the gist of which is explained in the following problem. Who will hear first a pianist's first note—a person in a concert hall seated some ten metres away from the piano, or a radio listener in his apartment some one hundred kilometres away from the hall? Strangely enough, the radio listener, even though he is ten thousand times farther away.

Indeed radio waves will travel 100 km in $\frac{100}{300,000} = \frac{1}{3,000}$ sec; meanwhile sound travels ten metres in $\frac{10}{340} = \frac{1}{34}$ sec. It follows that it is almost a hundred times faster to transmit sound by radio than to transmit sound through air.

Sound and Bullet

When the passengers in Jules Verne's projectile set off for the moon, they were mystified not to hear the roar of the tremendous gun that hurtled them into space. But it could not be otherwise. However deafening the noise, the speed with which it propagated—as any sound in air generally—was only 340 m/sec, whereas the projectile whizzed out with a speed of 11,000 m/sec. It was only natural that the sound of the shot did not reach the ears of the passengers, as their projectile was moving faster than sound.*

What about real projectiles and bullets? Do they move faster than sound? Or, on the contrary, does sound outrace them to warn the victim of the deadly projectile? Modern rifles impart to bullets a muzzle velocity nearly three times greater than the speed of sound in air—about 900 m/sec (the speed of sound at 0° is 332 m/sec). True, sound propagates uniformly, while a bullet slows down. However, over most of its trajectory a bullet will fly faster than sound. Hence, if in a skirmish you hear the sound of a shot or the whistle of a bullet, you needn't worry, the *bullet has already missed you*. The bullet will hit its victim before the sound of the shot would reach his ear.

False Explosion

The speed race between a flying object and its sound sometimes involuntarily compels us to draw absolutely erroneous conclusions. A curious illustration is afforded by a bolide, or a cannon ball, whizzing by high up above. The bolides that dive into our terrestrial atmosphere from outer space possess a tremen-

* Modern aircraft fly much faster than sound.—*Ed.*

dous velocity *dozens of times* greater than the speed of sound, even though they are retarded by atmospheric resistance.

As they cut through the air bolides often produce a noise greatly resembling peals of thunder. Suppose we are at point C (Fig. 152). High up above a bolide

Fig. 152

Imagined explosion of a bolide.

is rocketing along the trajectory AB. The sound it produced at point A will reach the ear at point C only when the bolide itself reaches point B. Since the bolide flies much faster than sound, it may reach a certain point D, from which we shall hear its sound, before we hear the sound it made at point A. Therefore we hear first the sound from point D and, only then, the sound from point A. Since the sound from point B reaches us also later than from point D, somewhere above us there should be a point K from which we should hear the bolide's noise soonest. All fond of mathematics will be able to figure out the exact position of this point, provided they determine the relation between the speeds of the bolide and the sound.

As a result what we *hear* will in no way resemble what we *see*. To the eye the bolide will appear first of all at point A from whence it will dash further along the trajectory AB. But to the ear the bolide will seem to be first somewhere above us at point K. Then we will hear simultaneously two sounds fading in opposite directions—from K to A, and from K to B. In other words, it will seem to us that the bolide had split into two parts, whizzing away in opposite directions.

Actually there was no explosion at all—which shows you how deceptive auditory impressions may be. Most likely many "eye witness" accounts of bolide explosions derive precisely from this auditory illusion.

If the Speed of Sound Were Less

If sound propagated in air with a speed much slower than 340 m/sec, auditory illusions would be far more frequent.

Suppose sound propagated with a speed not of 340 m/sec but of 340 mm/sec, which is slower than the speed with which one walks. Further suppose that you are sitting in an arm-chair and listening to a story being told by a friend having the habit of pacing the room as he talks. Ordinarily this would be no obstacle. But if the speed of sound were much less, you would not make head or tail of your friend's story. His initial utterances would be overtaken by more to mix and produce an incoherent jumble.

Incidentally, when your friend approached, his speech sounds would reach you in *reverse order*. At
first you would hear the sounds he had just uttered,

then sounds uttered a bit before, and after that sounds uttered still earlier, and so on and so forth. This would happen because the person uttering them would be outpacing them and uttering more sounds, meanwhile being all the time in front of the sounds just uttered.

The Slowest Conversation

But if you always thought the real speed of sound in air, a third of a kilometre a second, rapid enough, you will now have to change your mind. Suppose Moscow and Leningrad were linked up, instead of electric telephones, by an ordinary speaking tube, like the one on a steamship which the skipper uses to issue orders to the engine room. You are at the Leningrad end of this 650-km-long tube while your friend is at the Moscow end. You ask a question and wait for an answer. Some five minutes pass, then ten, fifteen, but there is still no reply. You are worried, and think something has happened. Your fears are unwarranted. Your question has *still not reached Moscow*. It is only half-way there. Another quarter of an hour will pass before your friend hears it. But since his reply will take the same half-hour to get from Moscow to Leningrad, you will hear your questions answered only one hour later.

This is easy to verify. It is 650 km from Leningrad to Moscow. Sound propagates with a speed of 1/3 km/sec. Conversing in this way from morning till night, you would barely manage to exchange a dozen sentences.*

. * The author has evidently deliberately disregarded the fading of sound with distance. You would never be able to conduct a conversation in this manner as the person at the other end of the tube would never hear you.—*Ed.*

There was a time, however, when even such a method of relaying news would have been considered very fast. A mere hundred years ago nobody dreamed even of electric telegraphs and telephones. If one then would have been able to transmit news over a distance of 650 km in the space of a few hours, it would have been considered ideal.

When Tsar Paul I was crowned, the story goes, the news of the coronation, which took place in Moscow, was relayed to the northern capital of St. Petersburg in the following manner. Soldiers were posted all along the road between the two capitals at a distance of 200 m apart. As soon as the cathedral bell chimed forth, the nearest soldier fired his gun into the air. As soon as the next soldier heard the shot, he fired his gun. The third did the same, and it thus took only three hours to relay the news to St. Petersburg, 650 km away.

If one could have heard the chiming of Moscow bells in St. Petersburg, this sound, as we already know, would have reached the northern capital half an hour later. This means that two and a half hours of the three hours required to convey the news went for each soldier to hear his neighbour's gun and make the necessary motions to fire his own gun. However short the delay, these thousands of tiny delays added up to two hours and a half.

The old optical telegraph was based on a similar principle. It transmitted light signals to the nearest post, which then relayed them further.

Tom-Tom Telegraph

The relaying of news by sound signals is current even today among tribes of Africa, Central America, and Polynesia. For this purpose special tom-toms, able to transmit sounds over great distances, are used. The signal is repeated in relays and very soon the population of an extensive area are "in the know" (Fig. 153).

When Italy was at war with Abyssinia, the Negus learned very quickly of all Italian troop movements. This puzzled the Italians who had no inkling of the

Fig. 153

A Fiji native using a tom-tom telegraph.

enemy's tom-toms. When Italy went to war with Abyssinia again, the general mobilisation order issued in Addys Ababa was "made public" in the same manner and relayed to the remotest villages in the space of a few hours.

Tom-toms were again used during the Anglo-Boer war. They enabled the Kaffirs to relay military information very quickly. The Capelanders learned everything several days before official dispatches came by messenger. Explorers claim that some African tribes have such an excellent system of sound signals, that they may be said to have a much better telegraph than the optical one used in Europe before the present electric one was introduced.

In one magazine I read the following about the tom-tom telegraph. R. Hasselden, the British Museum's archaeologist, was visiting the town of Ibada in the heart of Nigeria. The steady, dull rhythmic beat of tom-toms could be heard day and night. One morning the scientist heard the Negroes animatedly chattering among themselves. In response to his inquiries the sergeant told him that "a big ship of the white men has sunk and many white men have perished". Hasselden paid no heed to the rumour at the time. However, three days later he received a telegram (delayed because of disrupted communications) about the *Lusitania* disaster. Only then did he realise that the Negro grapevine was right and that it had "beat" its way in drum language from Cairo to Ibada. This was all the more surprising, since the tribes that had relayed the news spoke in totally different dialects while some were even at war with one another.

Acoustic Clouds and the Aerial Echo

Sound bounces back not only off solid obstacles but also off clouds. Even absolutely transparent air is able to reflect sound waves in certain circumstances, when for some reason it differs in sound-carrying capacity from the rest of the air mass. This is a phenomenon reminiscent of the "total reflection" in optics. The sound is repelled by an invisible barrier and we hear a mysterious echo whose source is undiscernible.

Tyndall discovered this curious phenomenon by accident, when experimenting with sound signals on the sea shore.

..."The echoes reached us," he wrote, "as if by magic, from the invisible acoustic clouds with which the optically transparent atmosphere was filled."

The famous British physicist gave the name of acoustic clouds to those areas of transparent air which reflected sound, thus producing an "aerial echo". This is what he has to say about it:

312

"Acoustic clouds, in fact, are incessantly floating or flying through the air. They have nothing whatever to do with ordinary clouds, fogs or haze. The most transparent atmosphere may be filled with them; converting days of extraordinary optical transparency into days of equally extraordinary acoustic opacity...

"The existence of these aerial echoes has been proved both by observation and experiment. They may arise either from air currents differently heated, or from air currents differently saturated with vapour."

Soundless Sound

Some people are deaf to such high-pitched sounds as the chirping of a cricket or the squeaking of a bat. Though not deaf generally and having normal auditory organs with no defects, they fail to hear high-pitched tones. Tyndall claims that there are people who can't even hear a sparrow chirruping.

Generally speaking, our ear does not catch every vibration occurring near by. We don't hear any sound when vibrations are less than 16 or more than 15,000-22,000 per second. The terminal threshold of audibility differs for different people, dropping to as low as 6,000 vibrations per second in the case of old persons. This is why one person will clearly hear a shrill high-pitched note while another won't.

Many insects, the mosquito and cricket, for instance, emit sounds having a pitch corresponding to 20,000 vibrations a second, which to some is audible and to others not. The latter, those insensitive to high-pitched notes, take pleasure in a quietness which for the former is an ear-racking world of shrill sounds. Tyndall had this observation to make about a stroll he and his friend once made in Switzerland:

"In *The Glaciers of the Alps* I have referred to a case of short auditory range, noticed by myself in crossing the Wengern Alps in company with a friend. The grass at each side of the path swarmed with insects, which to me rent the air with their shrill chirruping. My friend heard nothing of this, the insect-music lying beyond his limit of audition."

The squeak of a bat is one whole octave below the shrill shrieking of insects, as the vibrations produced have only half the frequency. But there are people whose borderline of audibility is still lower and to whom bats are noiseless creatures. On the contrary, dogs, as the famous Soviet physiologist Academician Pavlov experimentally demonstrated, hear tones up to 38,000 vibrations a second.

Ultrasounds in Technology

Modern physicists and engineers are able to produce "soundless sounds" with a frequency that is much greater than the sounds we have just mentioned. "Ultrasounds" may have vibrations ranging up to as many as 100,000,000,000,000 a second.

One way of producing ultrasonic vibrations is based on the property of plates cut in a certain manner from quartz crystals to electrify under compression (this is called piezoelectricity). This crystal will alternately contract and expand under the effect of periodic electric charges. In other words, it will vibrate and produce ultrasounds. The crystal is charged by a radio-tube generator with a frequency selected to conform with the so-called own period of the crystal's vibrations.*

---

*Since quartz crystals are expensive and feebly emit ultrasounds, they are hence used mostly in laboratories. Engineers have evolved for technical uses such synthetic materials as barium titanate ceramics.—*Ed.*

Though ultrasound is inaudible to us, it exhibits itself in other very manifest ways. If we immerse a vibrating plate in a jar of oil, a 10-cm "hump" will appear on the oil surface, due to ultrasonic sound, and drops of oil will spray up to 40 cm high. If we dip a metre-long glass rod in this oil bath, the hand holding the rod will be badly burned. The vibrating rod end will char a hole through a piece of wood, the energy of the ultrasounds having been transformed into thermal energy.

Scientists both in the Soviet Union and abroad are intensively investigating ultrasound. It strongly affects living organisms, causing the threads of seaweeds to snap, animal cells to burst, and blood cells to disintegrate. One or two minutes of ultrasound treatment is enough to kill small fish and frogs, and to raise an animal's body temperature to as muoh as 45°C—in the case of mice, for instance. The inaudible ultrasounds, like the invisible ultraviolet rays, are helping doctors to treat sick people.

Ultrasound is widely used in metallurgy, to detect impurities, blowholes, cavities and other inner defects. In ultrasonic fault detection, the metal being tested is greased with oil and subjected to ultrasonic vibrations. The defective area disperses the sound, casting what one might call a sound shadow, which appears in so clear a relief against the even ripples on the oil, that it can even be photographed.*

Ultrasound can be used to test for defects a bar of metal over a metre thick, which is way beyond the reach of X-rays. Moreover it will spot tiny defects of

---

* The method of ultrasonic fault detection was suggested by the Soviet scientist S.Ya. Sokolov back in 1928. The special ultrasonic vibration receivers used today dispense with oil and make the process simpler.—*Ed.*

as little as one millimetre in cross-section. There is no question that the future prospects are very promising.[*]

Giant's Bass and Midget's Treble

In the Soviet film *The New Gulliver* the Lilliputians speak in voices highly pitched to correspond with their tiny throats, while the boy-Gulliver, Petya, speaks in a deep voice. Nevertheless, when the picture was being made, it was grown-up actors who spoke for the Lilliputians and a real boy who spoke for Petya. How was the pitch changed? I was most surprised when film director Ptushko told me that the actors had not tried to change their voices at all. That was done for them by an original method based on the physical characteristics of sound.

To impart a high pitch to Lilliputian speech and a deep voice to Gulliver the actors speaking for the Lilliputians were recorded on a *slow* track while the boy playing Petya was recorded, on the contrary, on a *fast* track. The sound-film, however, was reproduced at the normal speed. You have probably guessed the outcome. Lilliputian speech is received by the audience when vibrations are more *intensive*, thus producing a *rise* in the pitch. Petya's speech, on the contrary, is reproduced at a *slower speed* than normal and, consequently, its tone should *drop*. As a result in *The New Gulliver*, the Lilliputians speak in voices a quint *higher* than that of a normal adult, while the boy-Gulliver speaks in a tone a quint *below* the normal.

---

[*] Ultrasounds exist in nature as well—in the noise of the wind and breakers at sea, for instance. Butterflies, cicadas and many other creatures both emit and receive ultrasounds. Bats use ultrasound when in flight as a sort of radar to circumvent obstacles.—*Ed.*

Thus was a slow-motion method used to produce a sound effect. One will often hear this, incidentally, when a gramophone record is played at a speed faster or slower than indicated.

Reading a Daily Twice a Day

The problem we are now going to discuss may seem, at first glance, to have nothing to do with either sound in particular, or physics in general. Still I would like you to pay heed to it, since it will facilitate understanding of what is to come later. In all likelihood you may have encountered another version of the same problem before.

Here it is. Daily, at noon, a train pulls out from Moscow for Vladivostok. Also daily, and again at noon, another train leaves Vladivostok for Moscow. Suppose the entire journey takes ten days. The question is: how many trains will you meet on your way from Vladivostok to Moscow? Most people rush to give the answer: ten. That's wrong though. Besides meeting on the way the ten trains that will leave Moscow after you depart from Vladivostok, you will also meet the trains already on the way at the time you left. Consequently, the right answer is not ten but twenty.

Let us now proceed further. Each Moscow train has on board fresh newspapers. If you are interested in Moscow news, you will naturally buy fresh papers at each stop. How many fresh newspapers then will you buy on your ten-day journey? I imagine that by now you will give the right answer of twenty. After all, each train you meet brings a fresh paper, and since you meet twenty, that means you read also twenty fresh papers. But since you will be travelling only ten days, *you will be reading a fresh daily twice a day.*

This seems a surprising paradox, doesn't it? And I don't suppose you would believe me unless you ever had a chance to put it to test in practice.

The Train Whistle Problem

If your ear for music is good, you have probably noticed how the pitch—not the loudness, but precisely the pitch—of the locomotive's whistle changes as an approaching train passes. While the two trains were drawing together, the pitch of the whistle was noticeably higher than after the trains had met and receded. If both trains are doing 50 km/h, the difference in the pitch reaches nearly a whole tone.

Why does this happen? The answer is easy to fathom, when you realise that the pitch depends on the frequency of vibrations per second and that you have an analogy with the newspaper problem mentioned above. The whistle of the approaching locomotive emits one and the same sound of a definite frequency. Your ear, however, receives a varying number of vibrations, depending on whether you are approaching, are stationary, or are receding.

In the same manner as on the train journey to Moscow, when you read daily newspaper twice a day, on approaching the source of sound you catch a vibration frequency of a rate *greater than the locomotive whistle's normal rate*. This, however, is no auditory illusion; it is your ear that receives the *increased number of vibrations* and you directly hear a tone of a *higher pitch*. As you move away, you get a smaller number of vibrations and hear a tone of a *lower pitch*.

If this explanation of mine has not satisfied you, try to trace in the mind's eye how the sound waves propagate from the locomotive whistle. Take the *stationary* locomotive (Fig. 154) to begin with. Its whist-

318

le emits wave-trains. For simplicity's sake we shall discuss only four wave-trains (see the upper wavy line). From the stationary locomotive they propagate in a definite time interval to one and the same distance in all directions. Wave 0 will reach observer *A*

Fig. 154

Train whistle problem. *A-B:* sound waves emitted by a stationary locomotive. *A′-B′:* sound waves emitted by a locomotive moving.

at the same time as observer *B*. Then both *A* and *B* will hear waves *1, 2, 3*, etc., simultaneously. The same number of waves strike the ears of both observers every second—which is why both hear one and the same tone.

It would be different were the whistling locomotive to be *moving* from *B* to *A* (the lower wavy line). Suppose at a certain moment the whistle would be at point *C′* and that while it had been emitting the four waves, it had managed to reach point *D*. Now compare the different propagation of the sound waves. Wave 0 from point *C′* will reach both observers *A′* and *B′* simultaneously. However, the fourth wave, emitted from point *D*, won't reach them simultaneously, as the distance *DA′* is less than *DB′*; consequently, it will reach A' before B'. The intermediate waves 1 and 2 will also arrive at B' later than at A', but there will be less difference in time. Consequently, A' will receive the sound waves *more often* than B' and will hear a tone of a *higher pitch*. At the same time, as the drawing clearly shows, the length of the waves moving towards A' will be corresponding shorter than the waves moving in the opposite direction, towards B'.

(The wavy lines of the drawing do not depict the *shape* of the sound waves. Air particles vibrate *longitudinally*, in the direction of the sound, and not transversely. The waves are given *transversely* solely to provide a graphic illustration. The crest of each wave corresponds to the maximum longitudinal contraction.)

The Doppler Effect

The phenomenon described above was discovered by the physicist Doppler and ever since has been known after him. It is also observed for light, as light likewise propagates in waves; the increasing frequency appears to the eye as a change in colour, while in the case of sound it is heard as a change in pitch.

The Doppler effect, as it is called, enables astronomers not only to detect whether a star is moving towards or from us, but also to estimate the velocity of this shift. In this case it is the side shift of the dark vertical lines in the spectrum that helps. A close examination of the direction of the shift and its extent, in the spectrum of a heavenly body, has enabled astronomers to make a whole series of astounding discoveries. Thus the Doppler effect has told us that the bright star of Sirius is receding by 75 km every second. It is so incredibly distant that its recession even by thousands of millions of kilometres would not appreciably affect its apparent brilliance, and most likely we would have

320

never guessed that it was moving away, if not for the Doppler effect.

This fact most glowingly shows physics to be a really all-embracing and *all-inclusive* science. Having evolved a law for *sound* waves, which are as much as several metres long, it then applies it to the infinitesimally small *light* waves, which are only a few ten-thousandths of a millimetre long, in order to calculate the rapid motion of huge suns in the fantastically distant reaches of outer space.

The Case of the Fine

When Doppler first concluded (in 1842) that the wavelengths of sound and light should change, as the source of emission approached or receded, he boldly postulated that this was precisely why stars were coloured. All stars, he conjectured, were actually white; however, many seemed coloured due to rapid motion in relation to us. Quickly approaching white stars, he reasoned, send us shortened light waves tinted green, blue, or violet. On the contrary, quickly receding white stars seem yellow or red.

Original but unquestionably wrong! For the eye to notice a change in colour due to motion, the stars would have to race with the enormous velocity of tens of thousands of kilometres a second—which even then wouldn't help, as at the same time the blue rays of an approaching white star would change to violet, the green ones would change to blue, while the violet ones would take over from the ultraviolets, and the red ones from the infrareds. In short, we should still have the same old components of white light and, despite the general shift in all the colours in the spectrum, our eye wouldn't notice any change in the general colour.

The shift of the dark lines in the spectra of stars

moving towards or from the observer is quite another thing. Sensitive instruments spot them, enabling us to determine the line-of-sight velocities of the stars. A good spectroscope will catch even a one-kilometre-per-second velocity.

The celebrated physicist Robert Wood recollected Doppler's error when the police were about to fine him for jumping a red traffic light. Wood, the story goes, assured the guardian of law and order that when driving fast you would see a red traffic light as a green one. Had the policeman been a bit of a physicist as well, he could have well told Wood that to see green in place of red, his car must do at least 135 million km/h—a totally incredible speed.

This is how it is calculated. Let l designate the wavelength of the light emitted by the traffic light, l' the wavelength received by Wood in his car, v the car's velocity, and c the velocity of light. Then we may write the following equation: $\frac{l}{l'} = 1 + \frac{v}{c}$. Knowing that the shortest, red, wavelength is 0.0063 mm, while the longest, green, wavelength is 0.0056 mm, and that the velocity of light is 300,000 km/sec, we get $\frac{0.0063}{0.0056} = 1 + \frac{v}{300,000}$, whence the speed of the car is $v = \frac{300,000}{8} = 37,500$ km/sec, or 135 million km/h. At this speed Wood, in the space of an hour or so, could have put a greater distance between himself and the police than from the earth to the sun. Still, the physicist was fined after all for "speeding".

With the Speed of Sound

What would you hear if you were to move away from a band with the speed of sound? You might think that since a mail-train passenger buys at all stations one and the same paper that was put out on the day of the train's departure, .by moving away with the speed of sound from the band, we would hear one and the same note that the orchestra had played when we had just started.

That would be wrong. Since you are moving away with the speed of sound, the sound waves emitted by the band would be in a state of rest with respect to you and would not impinge on your ear drums at all; you would hear nothing and think that the band had stopped playing.

Why did our comparison produce the wrong answer? Simply because we applied the analogy wrongly. After all the passenger could imagine—that is, if he forgot he was travelling—that ever since he had left Moscow, no fresh newspapers had been put out, since he was seeing one and the same newspaper sold all along the line. From his point of view the newspaper offices should have closed down—just as the band had seemed to have stopped playing if we were moving with the speed of sound.

Curiously enough, even scientists sometimes confuse the point, though it is really not so complicated as all that. I remember, when a schoolboy yet, having an argument with an astronomer who disagreed with my solution, and who claimed that when we moved away with the speed of sound we would hear one and the same tone all the time. The line of reasoning he chose was as follows:

"Suppose you have a note of a certain pitch struck," he wrote to me. "It has always had that sound and will always have it. A row of observers in space would hear it in succession and let us suppose, for the sake of argument, just as loudly. Why then shouldn't we hear it, were we able to hop with the speed of sound or even thought, to the side of any of these observers?"

In exactly the same way he argued that an observer moving away from lightning with the speed of light would always see this lightning.

"Imagine," he wrote, "an endless row of eyes in space. Each successive eye will see the flash in succession. So imagioe you visit each eye in succession: quite obviously you will see the flash of lightning all the time."

It goes without saying that neither assertion is right. In the conditions given we would neither hear the note nor see the lightning. The equation mentioned a bit earlier demonstrates that; for if $v = -c$ the wavelength l' will be infinite, which is all the same as saying that it isn't there.

* * *

We have come to the end of *Physics for Entertainment*. If, now that you have read it, you feel you would like to learn more about this boundless domain of knowledge from which this motley handful of simple facts has been culled, I shall consider my task fulfilled and, happily content, will write

The End

1. Can one see from a balloon how the earth rotates?

2. Will a weight, dropped from an airplane, fall vertically?

3. How can we get off moving railway trains with absolute safety?

4. When an icebreaker ploughs through ice, is the force it exerts equal to the ice's resistance?

5. Why does a rocket go up; and will it go up in a void?

6. Do any creatures move like rockets?

7. Does it always happen that forces exerted in different directions fail to make the objects to which they are applied move?

8. Why is an arched ceiling sturdier than a flat one?

9. How does the wind make a yacht go?

10. Would Archimedes have ever been able to lift the earth, even if he had found a point for a fulcrum?

11. Why does a knot hold?

12. Will knots help in the absence of friction?

13. How would we benefit and what would we lose, were there no friction?

14. Balance a broom on the back of a chair. Which part is heavier—the shorter or the longer part?

15. Why doesn't a spinning teetotum topple over?

16. When doesn't water pour out of an overturned glass?

17. When won't an unfettered ball roll down an incline?

18. Where is gravity greater—in London or Cairo?

19. Why do we never see the mutual attraction of objects in a room?

20. How far would you be able to jump on the moon?

21. How high up would a bullet go on the moon when fired vertically upwards with a muzzle velocity of 900 m/sec?

22. If you were to drop a weight into a shaft going right through the centre of the earth along its diameter, would it stop in the absence of air resistance?

23. How should one dig a tunnel through a mountain to avoid its flooding by rain?

24. Can you hurtle something into space so that it never comes back?

25. Where would the nonswimmer never sink?

26. How does an icebreaker go through ice?

27. Do shipwrecks reach the ocean bed?

28. On what physical law is the lifting of sunken ships based?

29. What is the tub problem and do arithmetic books provide the correct solution?

30. Can we make water flow out of a vessel in a uniform jet?

31. Would two harnesses of 8 elephants apiece—in place of the two harnesses of 8 horses—have pulled the Magdeburg hemispheres apart? (An elephant is presumably five times stronger than a horse.)

32. What is the working principle of an atomiser?

33. Why do two parallel ships attract each other?

34. What is the role of a fish's bladder?

35. What are the two different fluid flows in physics?

36. Why does smoke curl when coming out of a chimney?

37. Why does a flag flutter in the breeze?

38. Why does desert sand "ripple"?

39. How high up must one go for atmospheric pressure to drop by one-thousandth?

40. Is Mariotte's law applicable to air at a pressure of 500 atmospheres?

41. Is the temperature a thermometer shows in windy weather lower than that in windless weather?

42. Why is a frost worse in windy weather than in quiet weather?

43. Will a wind always refresh us in hot weather?

44. What is the effect of coolers based on?

45. Can you make a cooler without ice?

46. Can you bear a 100°C heat?

47. Why is it easier to bear a 36°C heat wave in Central Asia than a 24°C heat wave in Leningrad?

48. What is the function of glass in the paraffin lamp?

49. Why don't the products of combustion extinguish the flame of a paraffin lamp or a candle?

50. How would a flame burn were there no gravity?

51. How would water heat up on a primus stove were there no gravity?

52. Why does water put fire out?

53. How can we fight fire with fire?

54. Will pure water ever boil in a vessel heated up by boiling water?

55. Will water ever freeze in a bottle immersed in a mixture of water and ice?

56. Can we get water to boil at room temperature?

57. How can we determine atmospheric pressure by using a thermometer?

58. Is there anything like hot ice?

59. Which kind of magnet—natural or man-made—would be more powerful?

60. What other metals besides iron can a magnet attract?

61. Are there any metals which a powerful magnet would be able to repel?

62. Will a magnet affect liquids or gases?

63. Where in the world would a compass needle point North with both ends?

64. What attracts more strongly—the iron the magnet, or the magnet the iron?

65. Which sense organ is susceptible to magnetic forces?

66. Can an electromagnetic crane hoist molten metal?

67. Why are powerful magnets dangerous for gold watches?

68. What is a radium clock?

69. How do we determine the age of the earth and minerals by radioactive decay?

70. Why can birds perch on electric wires with impunity?

71. How long does lightning last?

72. At what angle should we put two mirrors to get seven reflections?

73. What difference is there between a sun-powered motor and a sun-powered heater?

74. What is "helio-engineering"?

75. Why is the crystalline lens of a fish's eye spherical in shape?

76. Can you read a book with your head under water?

77. Which of the two—a diver in a helmet or a person without goggles—will see better under water?

78. Can we make a biconcave lens magnify and vice versa?

79. Why does the eye see the bottom of a pond as elevated?

80. What is the critical angle?

81. What is total reflection?

82. Do the fish's silvery scales help it in any way?

83. What is the blind spot and how can we find it?

84. What is the angle of vision?

85. How far away should we hold a six-penny bit for it to hide the full moon?

86. How far apart are the sides of a 1′ angle 10 metres away from the apex?

87. Jupiter's diameter is roughly ten times greater than that of the earth. How far away is Jupiter when its disc is observed at an angle of 40″?

88. How should we understand the expressions: "a microscope makes things 300 times bigger" or "a telescope brings things 500 times closer"?

89. Why do motor-car wheels often seem to be revolving in the opposite direction on a screen?

90. Can we get a quickly spinning object to appear as a stationary one to the eye?

91. Can a rabbit see everything going on around it without turning its head?

92. Is it true that all cats are grey when the candles are out?

93. What propagates faster: a radio signal or sound in air?

94. What moves faster: a rifle bullet or the sound of the shot?

95. What sound vibrations are we unable to hear?

96. Have engineers been able to put "soundless sounds" to any use?

97. What is an "acoustic cloud"?

98. How does the pitch of a whistle of an approaching locomotive change?

99. What would we hear if we were to move away from a band with the speed of sound?